Michael Seifert

Hidden Markov Models with Applications in Computational Biology

Michael Seifert

Hidden Markov Models with Applications in Computational Biology

Model Extensions and Advanced Analysis of DNA Microarray Data

Südwestdeutscher Verlag für Hochschulschriften

Impressum / Imprint

Bibliografische Information der Deutschen Nationalbibliothek: Die Deutsche Nationalbibliothek verzeichnet diese Publikation in der Deutschen Nationalbibliografie; detaillierte bibliografische Daten sind im Internet über http://dnb.d-nb.de abrufbar.

Alle in diesem Buch genannten Marken und Produktnamen unterliegen warenzeichen-, marken- oder patentrechtlichem Schutz bzw. sind Warenzeichen oder eingetragene Warenzeichen der jeweiligen Inhaber. Die Wiedergabe von Marken, Produktnamen, Gebrauchsnamen, Handelsnamen, Warenbezeichnungen u.s.w. in diesem Werk berechtigt auch ohne besondere Kennzeichnung nicht zu der Annahme, dass solche Namen im Sinne der Warenzeichen- und Markenschutzgesetzgebung als frei zu betrachten wären und daher von jedermann benutzt werden dürften.

Bibliographic information published by the Deutsche Nationalbibliothek: The Deutsche Nationalbibliothek lists this publication in the Deutsche Nationalbibliografie; detailed bibliographic data are available in the Internet at http://dnb.d-nb.de.

Any brand names and product names mentioned in this book are subject to trademark, brand or patent protection and are trademarks or registered trademarks of their respective holders. The use of brand names, product names, common names, trade names, product descriptions etc. even without a particular marking in this works is in no way to be construed to mean that such names may be regarded as unrestricted in respect of trademark and brand protection legislation and could thus be used by anyone.

Coverbild / Cover image: www.ingimage.com

Verlag / Publisher:
Südwestdeutscher Verlag für Hochschulschriften
ist ein Imprint der / is a trademark of
AV Akademikerverlag GmbH & Co. KG
Heinrich-Böcking-Str. 6-8, 66121 Saarbrücken, Deutschland / Germany
Email: info@svh-verlag.de

Herstellung: siehe letzte Seite /
Printed at: see last page
ISBN: 978-3-8381-3604-2

Zugl. / Approved by: Halle, Martin Luther University Halle-Wittenberg, Diss., 2010

Copyright © 2013 AV Akademikerverlag GmbH & Co. KG
Alle Rechte vorbehalten. / All rights reserved. Saarbrücken 2013

To my family

Preface

Standard first-order Hidden Markov Models (*HMM*s) are frequently used methods for the analysis of sequential data in a broad range of scientific domains including applications in speech recognition or computational biology. *HMM*s are versatile and structurally simple models enabling probabilistic modeling based on a sound mathematical and algorithmic grounding. However, still most of the developed *HMM*-based approaches are only applying concepts of standard first-order *HMM*s. This is sufficient enough to reach good results in many applications, but most results can also be substantially improved by utilizing higher-order *HMM*s as demonstrated in the domains of speech or handwriting recognition, robotics or the analysis of protein and DNA sequences.

My main intention for writing this thesis was to create an easily accessible and comprehensive extension of the mathematical and algorithmic basics of standard first-order *HMM*s to higher-order *HMM*s coupled with some selected practical applications in modern computational biology. Additionally, one of my other goals was to improve and ease the creation of biologically meaningful models by the integration of biological prior knowledge into the training of *HMM*s. This has indeed proved to be of utmost importance for my selected case studies. Moreover, I have also developed two novel more specialized model extensions (i) parsimonious higher-order *HMM*s and (ii) *HMM*s with scaled transition matrices to improve the modeling of sequential data. The parsimonious higher-order *HMM* enables a data-dependent interpolation between a mixture model that ignores spatial dependencies between adjacent sequence positions and a higher-order *HMM* that exhaustively models spatial dependencies in a sequence of data. This allows to use improved modeling characteristics of higher-order *HMM*s in cases of limited availability of sequential data and contributes to avoid overfitting of models. The *HMM* with scaled transition matrices enables the integration of additional prior knowledge into the state-transition process of the model for realizing an improved modeling of measurements in chromosomal contexts such as the differentiation between gene-pair orientations or the modeling of the distance between adjacent genes on a chromosome. I have applied all these different models in specific contexts covering the identification of differentially expressed genes in breast tumors, the identification of transcription factor target genes in yeast and plant data and for comparative genomics of two accessions of the model plant *Arabidopsis thaliana*. Additionally, I have made use of independent data for model evaluations in each case study and I have compared all extensions of *HMM*s to standard first-order *HMM*s and other typically used methods.

With my thesis, I try to address readers that already have some basic knowledge about standard first-order *HMM*s. What the reader can expect is to gain more insights into higher-order *HMM*s followed by details how one can derive specific model extensions. This should also allow the reader to develop own approaches based on higher-order *HMM*s. Since my research interests are in computational biology, some basic knowledge in biology is helpful, but definitely not mandatory for getting into the algorithmic extensions.

I have structured my thesis into two main parts. After a general introduction, the thesis starts with a theoretical part (chapters 2 – 4) where I develop the algorithmic basics of higher-order *HMM*s and where I describe the extensions to parsimonious higher-order *HMM*s and *HMM*s with scaled transition matrices. In the second part (chapters 5 – 8), I consider applications where I apply these models to the analysis of different DNA microarray data sets. Finally, I close my thesis with a comprehensive discussion of the achieved results. The first part of the thesis can be read independently of the second part and can be considered as a repository for the algorithmic basics of higher-order *HMM*s that has not been outlined in such a great detail in the existing literature.

My thesis has of course also limitations where I'm not going as deep into the individual subjects as I would have loved to do without having time and project-related financial constraints. For example, I did not consequently extend the *HMM*s with scaled transition matrices to higher-order *HMM*s. That is relatively straightforward and may lead to an additional improvement in the identification of differentially expressed genes in tumor. Additionally, I also did not perform the comparisons to existing methods for the case studies on the breast cancer gene expression data and on the comparative genomics of the model plant *Arabidopsis thaliana* in such great detail as I have recently done this in the corresponding manuscripts Seifert et al. (2011) and Seifert et al. (2012). I'm also not providing an exhaustive review of existing literature of higher-order *HMM*s. The interested reader may have a look at Seifert et al. (2012) and references therein as a starting point. I also do not discuss research on higher-order *HMM*s by du Preez (1998) in the field of speech recognition or by Wang (2006) in the field of motion capturing. Adapted versions of their proposed strategies may also be of value in computational biology.

Additionally, I also have to note that the present version of this thesis represents a revision of my initial thesis (Seifert (2010)). I worked on this thesis from June 2006 up to April 2010 at the IPK in Gatersleben. Here, I have extended my initial thesis by additional sections for further reading at the end of each case study (chapters 5 – 8). In this additional sections, I cover the progress that has been made since April 2010 and I provide links to publicly available implementations of the developed *HMM*-based methods and used data sets. I have also changed the original title from 'Extensions of Hidden Markov Models for the analysis of DNA microarray data' to 'Hidden Markov Models with Applications in Computational Biology – Model Extensions and Advanced Analysis of DNA Microarray Data' to broaden the spectrum of potentially interested readers and to better account for the potential general applicability of the developed approaches. Additionally, access to colored figures and ongoing research activities are provided under: https://sites.google.com/site/michaelseiferthmm/hmm-book.

My research on *HMM*s would not have been possible without the support from different sides. I'm grateful to Marc Strickert and Ivo Grosse for offering the opportunity to work on this topic in their research groups. My fascination for *HMM*s has been initiated by Alexander Schliep, who provided me the possibility to gain more insights on *HMM*s during a very inspiring time when I was writing my diploma thesis. The development of parsimonious higher-order *HMM*s would not have been possible without André Gohr providing an implementation of the parsimonious cluster algorithm that is sitting at the heart of these models. Realizations of publicly available implementations of my different *HMM*-based approaches would have been much harder to realize without the great support by Jens Keilwagen and Jan Grau utilizing their Jstacs framework (Grau et al. (2012)). I would not have been able to perform all my different case studies without wet-lab support and many valuable discussions. Gudrun Mönke, Urs Hähnel, Helmut Bäumlein, Lothar Altschmied Udo Conrad provided the ABI3 ChIP-chip data and performed validations of ABI3 target genes. I'm grateful to Ali Mohammad Banaei Moghadam, Andreas Houben, Michael Florian Mette, François Roudier and Vincent Colot for providing the Arabidopsis Array-CGH data. I would not have been able to write this thesis without the financial support from the BMBF (0312706A) and the Ministry of Culture Saxony-Anhalt (XP3624HP/0606T). I'm also grateful to the journals of *Bioinformatics* and *PLoS Computational Biology* for enabling me to reuse materials of my publications that I have written based on my initial thesis. Last but not least, I want to thank my family for the support in all different stages of writing this thesis.

Dresden, January 2013

Michael Seifert

Abstract

Hidden Markov Models (*HMM*s) are very popular tools in the field of computational biology for the analysis of sequential data from genomic studies. In this field, currently almost all *HMM*-based approaches make use of the theory behind standard first-order *HMM*s that model dependencies between directly adjacent positions in a sequence of data. This modeling of first-order dependencies can be a limitation for the analysis of sequential data. To overcome this, the algorithmic basics of first-order *HMM*s are comprehensively extended to higher-order *HMM*s for realizing dependencies between a position and corresponding most recent predecessor positions in a sequence of data. An important part of these extensions establishes the basis to integrate biological prior knowledge about a data set into the training of *HMM*s by using the Bayesian Baum-Welch algorithm. The goal of making use of such information is to improve the realization of biologically meaningful models. In addition to this, genomic features comprising the distance between adjacent genes on a chromosome or the orientation of adjacent genes to each other are modeled by a specifically developed *HMM* with scaled transition matrices. This model is applied to the extended analysis of recent high-throughput DNA microarray data sets of different organisms. In the context of the analysis of human gene expression data, *HMM*s with scaled transition matrices are used to model chromosomal distances of adjacent genes for improving the identification of differentially expressed genes in breast tumors. For the yeast *Saccharomyces cerevisiae* and the model plant *Arabidopsis thaliana*, *HMM*s with scaled transition matrices that distinguish between orientations of adjacent genes on a chromosome are applied to refine the prediction of transcription factor target genes from ChIP-chip[1] data. Besides this extended *HMM*, another extension, the parsimonious higher-order *HMM*, is developed based on the theory behind higher-order *HMM*s. The parsimonious higher-order *HMM* reduces the huge number of free transition parameters of a higher-order *HMM* in a data-dependent manner. Both, higher-order and parsimonious higher-order *HMM*s are applied to recent Array-CGH[2] data for predicting sequence polymorphisms in the genomes of two important accessions of *Arabidopsis thaliana*. Generally, all extensions of *HMM*s are compared to standard first-order *HMM*s and other typically used methods. Moreover, the predictions of all methods are comprehensively validated by making use of published literature, specific data bases, comparison to other technologies, or additional wet-lab experiments.

---

[1] ChIP-chip: Chromatin-immunoprecipitation coupled with hybridization on a DNA microarray (chip)
[2] Array-CGH: Comparative genomic hybridization on a DNA microarray

Contents

1 **Introduction**   1

2 **Markov Models**   6
   2.1 First-Order Markov Models 7
     2.1.1 Homogeneous First-Order Markov Model 7
     2.1.2 Inhomogeneous First-Order Markov Model 8
   2.2 Higher-Order Markov Models 11
     2.2.1 Homogeneous Higher-Order Markov Model 12
     2.2.2 Inhomogeneous Higher-Order Markov Model 13

3 **Hidden Markov Models**   15
   3.1 Homogeneous First-Order Hidden Markov Models 17
   3.2 Inhomogeneous Higher-Order Hidden Markov Model 19
   3.3 Solving the Likelihood Problem 20
     3.3.1 Forward Algorithm 22
     3.3.2 Backward Algorithm 24
     3.3.3 Forward-Backward Procedure 26
   3.4 Solving the Optimal State Sequence Problem 26
     3.4.1 State-Posterior Algorithm 26
     3.4.2 Viterbi Algorithm 27
   3.5 Solving the Maximum Likelihood Problem 31
     3.5.1 Baum-Welch Algorithm 31
     3.5.2 Separating Baum's Auxiliary Function Into Parameter Classes . 34
     3.5.3 Estimating HHMM Parameters 38
     3.5.4 Computational Scheme of the Baum-Welch Algorithm 40
   3.6 Prior 41
     3.6.1 Initial State Parameter Prior 42
     3.6.2 Transition Parameter Prior 42
     3.6.3 Emission Parameter Prior 43

Contents

- 3.7 Solving the Maximum A Posteriori Problem 43
  - 3.7.1 Bayesian Baum-Welch Algorithm 44
  - 3.7.2 Estimating HHMM Parameters 46
  - 3.7.3 Computational Scheme of the Bayesian Baum-Welch Algorithm . 50

4 Parsimonious Higher-Order Hidden Markov Models   51
- 4.1 Partitions of the Set of Hidden States 52
  - 4.1.1 Computing the Partitions 52
  - 4.1.2 Number of Partitions 53
  - 4.1.3 Set of Partitions 54
- 4.2 Tree-based Representation of State Contexts 54
- 4.3 Inhomogeneous Parsimonious Higher-Order Hidden Markov Model ... 57
- 4.4 Solving the Maximum A Posteriori Problem 58
  - 4.4.1 Tree-Based Baum's Auxiliary Function for Transition Parameters 59
  - 4.4.2 Tree-Based Transition Prior 60
  - 4.4.3 Tree Structure Prior 60
  - 4.4.4 Bayesian Baum-Welch Algorithm 61
  - 4.4.5 Scoring Scheme for Tree Structures 62
  - 4.4.6 Estimating Transition Parameters for an Equivalence Class ... 64
  - 4.4.7 Basics for Determining Optimal Tree Structures and Corresponding Transition Parameters 65
  - 4.4.8 Extended Tree 66
  - 4.4.9 Parsimonious Cluster Algorithm 67
  - 4.4.10 Computational Complexity of the Parsimonious Cluster Algorithm 69

5 Hidden Markov Models with Scaled Transition Matrices   71
- 5.1 Scaling of Transition Matrices 72
- 5.2 Hidden Markov Model with Scaled Transition Matrices 73
- 5.3 Solving the Maximum A Posteriori Problem 74
  - 5.3.1 Transition Prior 75
  - 5.3.2 Baum's auxiliary function for Transition Parameters 75
  - 5.3.3 Estimation of Transition Parameters 76

6 Analysis of Breast Cancer Gene Expression Data   80
- 6.1 Breast Cancer Gene Expression Data Set 82
- 6.2 Methods for Breast Cancer Gene Expression Data Analysis 83

		6.2.1	Hidden Markov Model approach	83

      6.2.1   Hidden Markov Model approach 83
      6.2.2   Hidden Markov Model with two scaled transition matrices 86
      6.2.3   Related approaches from the field of Array-CGH analysis 88
  6.3   Breast Cancer Gene Expression Data Analysis 89
      6.3.1   Comparison of Baum-Welch and Bayesian Baum-Welch training 89
      6.3.2   Comparison of HMM, SHMM, and related approaches 90
      6.3.3   Effect of chromosomal distances of genes on self-transition probabilities of SHMMs 92
      6.3.4   Validation of prediction results of HMM, SHMMs, and GLAD . . . 94
      6.3.5   Influence of modeling chromosomal locations and distances of genes on the prediction results 96
      6.3.6   Hotspots of under-expression and over-expression 98
  6.4   Further reading 101

7 Analysis of Promoter Array ChIP-chip Data   102
  7.1   Promoter Array Data Sets 105
      7.1.1   Yeast Data Set 105
      7.1.2   Arabidopsis Data Set 105
  7.2   Methods for Promoter Array Data Analysis 106
      7.2.1   Standard Log-Fold-Change analysis 106
      7.2.2   Basic first-order Hidden Markov Model 106
      7.2.3   Hidden Markov Model with two scaled transition matrices 107
  7.3   Identification of Common Target Genes of Yeast Cell Cycle Regulators . 109
      7.3.1   Prediction of putative common target genes 109
      7.3.2   Validation of putative common target genes 110
  7.4   Identification of Arabidopsis ABI3 Target Genes 111
      7.4.1   Systematic analysis of differences between HMM and SHMM . . 112
      7.4.2   Comparison of ABI3 target gene predictions of LFC, HMM, and SHMM 113
      7.4.3   Biological validation of putative ABI3 target genes 115
  7.5   Further reading 117

8 Analysis of Arabidopsis Array-CGH Data   118
  8.1   Arabidopsis Array-CGH Data Set 121
  8.2   Methods for Array-CGH Data Analysis 121
      8.2.1   Hidden Markov Model approaches 121

　　　　8.2.2　Related approaches for the analysis of Array-CGH data 124
　　8.3　Arabidopsis Array-CGH Data Analysis 124
　　　　8.3.1　Analysis of dependencies between log-ratios 124
　　　　8.3.2　SOLiD and Affymetrix resequencing data for validating the Array-CGH data set 125
　　　　8.3.3　Performance of HHMMs on the Array-CGH data set 129
　　　　8.3.4　Performance of PHHMMs on the Array-CGH data set 130
　　　　8.3.5　Selected tree structures of PHHMMs 134
　　　　8.3.6　Comparison of PHHMMs and HHMMs at a higher FPR 134
　　　　8.3.7　Comparison of PHHMMs to other methods 136
　　　　8.3.8　Analysis of PHHMM predictions in the context of the genome annotation 139
　　8.4　Further reading 142

9 Conclusions   143

Bibliography   150

Index   161

1 Introduction

Hidden Markov Models (*HMM*s) are probabilistic models for the analysis of sequential data. The theory behind these models has initially been developed and studied in the late 1960s and early 1970s in a series of papers (Rabiner (1989)). First practical applications of *HMM*s have been published in the domain of speech recognition (Rabiner (1989); Juang and Rabiner (1991)). Strengths of *HMM*s include their sound mathematical grounding and the availability of efficient algorithms for sequential data analysis. *HMM*s can be visualized by mathematical graphs that easily enable the design of a model for a specific application. All these characteristics have contributed to the broad usage of *HMM*s as one of the basic models in applied sciences (Mac Donald and Zucchini (1997); Durbin et al. (1998); Jelinek (1998)). In the field of speech recognition, *HMM*s are used to classify spoken words, digits, or even more complex speech signals (Rabiner (1989); Juang and Rabiner (1991); Jelinek (1998)). More recent applications in this field use *HMM*s for the separation of speech and music (Ajmera et al. (2002)), or for the automatic recognition of human emotions from speech signals (Schuller et al. (2003)). Other applications of *HMM*s can be found in the field of image segmentation (Li and Gray (2000)). This includes the usage of *HMM*s for the segmentation of radar images taken by satellites (Derrode et al. (2004)) as well as the classification of images into land-usage categories (Mari and Le Ber (2006)).
About two decades ago first applications of *HMM*s have found their way into computational biology. Based on these models, human genetic linkage maps have been constructed by Lander et al. (1987), and the compositional structure of DNA sequences has been analyzed by Churchill (1989). Over the years, several applications of *HMM*s have been identified including the analysis of DNA and protein sequences as well as the analysis of DNA microarray data. In the context of DNA and protein sequence data analysis, applications comprise gene finding (Kulp et al. (1996); Krogh (1997)), pairwise sequence alignments (Durbin et al. (1998)), homology searches (Krogh (1994); Eddy (1998)), and the characterization of protein structures (Campoux et al. (1999); Bystroff et al. (2000)). An overview of these applications is provided by the textbook of Durbin et al. (1998) and by the reviews of Cherry (2001) and de Fonzo et al. (2007).

1. Introduction

In the last decade, the DNA microarray technology has been developed to a powerful platform for the functional analysis of genomes. Applications include the analysis of gene expression profiles (Duggan et al. (1999); Lipshutz et al. (1999); Schulze et al. (2001)), the prediction of target regions of DNA-binding proteins like transcription factors or histones (Ren et al. (2000); Iyer et al. (2001); Martienssen et al. (2005)), and the study of sequence polymorphisms like deletions or amplifications of DNA segments between genomes (Mantripragada et al. (2004); Pinkel and Albertson (2005); Clark et al. (2007)). The DNA microarray technology provides the opportunity to analyse thousands of features (oligonucleotides or longer single-stranded DNA fragments) simultaneously within a single biological experiment. The features are located on a glass slide at high density to identify a complex mixture of target molecules (Ekins and Chu (1999)). In most studies, these target molecules represent either DNA or RNA isolated from cells or tissues (Hoheisel (2006)). Every feature has the capacity to be recognized by its complementary target sequence through base pairing, and the labeling of this complementary sequence by radioactively marked DNA bases or fluorescent dyes enables the quantification of the amount of complementary sequences that bound to specific features (Lipshutz et al. (1999); Duggan et al. (1999)). A broad overview of latest technologies and applications of microarrays is given in the recent reviews by Hoheisel (2006) or Shiu and Borevitz (2008).

In this thesis, DNA microarray data sets provided through cooperations at the IPK Gatersleben and DNA microarray data sets available from public sources are analyzed. In the domain of gene expression microarray data, the breast cancer data set published by Pollack et al. (2002) is considered. To study target regions of DNA-binding proteins, ChIP-chip (chromatin immunoprecipitation coupled with hybridization to a DNA microarray; see Ren et al. (2000) or Iyer et al. (2001)) experiments by Lee et al. (2002) are used to identify target genes of yeast cell cycle transcription factors. As part of a cooperation in the project Arabido-Seed (2006-2009), ChIP-chip data of the seed-specific transcription factor ABI3 of the model plant *Arabidopsis thaliana* is studied. For the analysis of genomic differences between two genomes, Array-CGH (comparative genomic hybridization on a DNA microarray; see Martienssen et al. (2005)) data of two accessions of *A. thaliana* available through the cooperation with A. Banaei (2008-2009) is investigated.

Generally, the huge number of measurements and the low number of replicates of a biological experiment put great challenges on the development of methods for the analysis of DNA microarray data (Piatetsky-Shapiro and Tamayo (2003)). In this context,

*HMM*s have been applied successfully to the clustering of gene expression time course data (Schliep et al. (2003, 2004); Yuan and Kendziorski (2006)), to the genome-wide prediction of DNA target regions of transcription factors or histones from ChIP-chip data (Li et al. (2005); Ji and Wong (2005); Humburg et al. (2008)), or for the identification of deletions and amplifications of DNA regions in Array-CGH data of tumors (Fridlyand et al. (2004); Marioni et al. (2006); Rueda and Diaz-Uriarte (2007)). Most of these *HMM* approaches are based on the standard first-order *HMM* that has been reviewed by Rabiner (1989). First extensions of this *HMM* are realized by Marioni et al. (2006) and Rueda and Diaz-Uriarte (2007) through the integration of chromosomal distances of measured features to improve the analysis of Array-CGH data.

In general, additional genomic information like chromosomal locations, distances or chromosomal orientations of features represented on a DNA microarray could be useful to enhance the *HMM*-based data analysis. Besides this, only little attention has currently been paid on the integration of biological prior knowledge about an experiment. This prior knowledge includes information like under-expressed genes are expected to have lower expression levels than unchanged expressed genes in tumor, a DNA target region bound by a transcription factor is expected to have greater measurements than a non-target region, as well as amplifications of DNA segments are expected to have greater measurements than unchanged DNA segments. The modeling of such biological prior knowledge by an *HMM* could improve the specification of a biologically meaningful model for the analysis of DNA microarray data. In addition to this, so far, no attention has been given to *HMM*s that model higher-order dependencies between measured features in the context of their chromosomal locations. The standard first-order *HMM* only realizes dependencies between a feature and its directly adjacent feature on a chromosome. Thus, the extension of the theory behind a first-order *HMM* to a higher-order *HMM* that models dependencies between a feature and its most recent predecessor features could provide further improvements for the analysis of DNA microarray data.

Published applications of higher-order *HMM*s are very rare except that like de Villiers and du Preez (2001) and Lee and Lee (2006) in the domain of speech recognition, those by Mari and Le Ber (2006) and Benyoussef et al. (2008) in the domain of image segmentation, or the study by Ching et al. (2003) for the modeling of DNA sequences. Still, a comprehensive introduction to the algorithmic basics of higher-order *HMM*s is currently not available. For that reason, the first main objective of this thesis is the general development of the algorithmic basics of higher-order *HMM*s. This includes

1. Introduction

the integration of additional genomic features as well as the ability to model biological prior knowledge. The second main objective of this thesis is to develop two specific extensions of *HMMs* with respect to the algorithmic basics of the higher-order *HMMs*. The first extension represents a parsimonious higher-order *HMM* that integrates a specific algorithmic concept that has been introduced by Bourguignon and Robelin (2004) and which has been later refined by Gohr (2006). The second extension considers an *HMM* with scaled transition matrices that has initially been described in Seifert (2006) for including additional genomic information into the analysis of gene expression data. This model is now extended to enable the integration of biological prior knowledge. The third main objective of this thesis is to apply the developed *HMMs* to the previously described data sets to demonstrate the broad usability of specific *HMMs* on specific types of data sets. This leads directly to the fourth objective of this thesis, the validation of the analysis results obtained from specific *HMMs* using independent validation data like biological validation experiments done by biologists, comparisons to published data sets, or by analyzing published literature and specific data bases. The general data analysis pipeline used in this thesis is shown in Fig. 1.1.

In summary, this thesis comprises a theoretical part (chapters 2 – 4) in which the theory behind standard *HMMs* is extended and an application part (chapters 5 – 8) in which the developed models are used to analyze DNA microarray data of a broad range of current research directions. Chapter 2 gives an overview of Markov Models to provide the basics for *HMMs*. In chapter 3 the transition from Markov Models to *HMMs* is motivated and the algorithmic basics of higher-order *HMMs* are developed. In chapter 4 the parsimonious higher-order *HMM* is introduced. Chapter 5 considers the extension of the *HMMs* with scaled transition matrices to enable the integration of biological prior knowledge. In chapter 6 breast cancer gene expression data is analyzed by making use of chromosomal distances between genes. Chapter 7 comprises the analysis of ChIP-chip data of the yeast *S. cerevisiae* and of the model plant *A. thaliana* by utilizing gene pair orientations on DNA. In chapter 8 genomic differences between two accessions of *A. thaliana* are determined based on Array-CGH data. Finally, general conclusions and an outlook to possible future directions are given in chapter 9.

1. Introduction

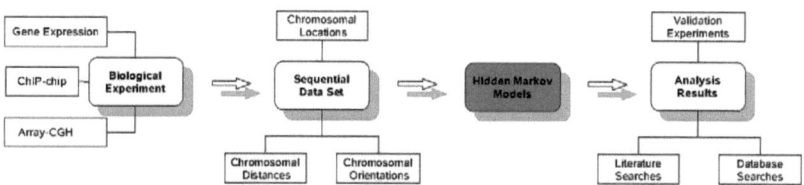

Figure 1.1: General overview of the DNA microarray data analysis pipeline applied in this thesis. Data generated in a biological experiment is organized as a sequential data set by making use of additional genomic information. The sequential data set is analyzed by specific *HMMs* developed in this thesis. The analysis results are further validated by independent data sources.

2 Markov Models

The Markov Model (*MM*) is a probabilistic model for representing dependencies in sequential data. The theory behind this model goes back to the Russian mathematician Andrej A. Markov (1856-1922). In many disciplines, including physics, chemistry, and biology, *MM*s have been applied successfully to represent temporal and spatial sequences (Berchtold and Raftery (2002)). The textbook by Durbin et al. (1998) includes a good introduction to *MM*s for the analysis of DNA sequence data. A more general introduction to *MM*s can be found in the textbook of Bishop (2006). Generally, a *MM* is used to model statistical dependencies between consecutive data points in a sequence of data points. Different *MM*s can be considered to model these dependencies by making assumptions about the number of consecutive data points that have an effect on the next data point. First-order *MM*s only model dependencies between a current data point and its next data point, while higher-order *MM*s extend this by modeling dependencies between a specific number of predecessor data points of a next data point (Berchtold and Raftery (2002)). Additional information about the data can be integrated into these modeling assumptions to improve the modeling of sequential data. In this chapter, *MM*s and motivated extensions are outlined to provide the basics of this thesis.

Goals of this Chapter

1. Homogeneous first-order *MM*s are briefly introduced and extended to inhomogeneous first-order *MM*s that integrate additional information into the state-transition process.

2. Homogeneous first-order *MM*s are extended to homogeneous higher-order *MM*s. Based on this, homogeneous higher-order *MM*s are extended to inhomogeneous higher-order *MM*s.

2.1 First-Order Markov Models

The basis for the modeling of sequential data by a *MM* can be derived from the joint distribution

$$P[\vec{Q}] = P[Q_1] \cdot \prod_{t=1}^{T-1} P[Q_{t+1} | Q_t, \ldots, Q_1] \tag{2.1}$$

for a sequence of discrete random variables $\vec{Q} := (Q_1, \ldots, Q_T)$. Each discrete random variable Q_t is taking values in the finite set of states $S := \{S_1, \ldots, S_N\}$. The joint distribution in (2.1) models statistical dependencies between each next random variable Q_{t+1} and all its predecessors $Q_t, \ldots, Q_1$. These long-range dependencies can be relaxed by making a specific assumption about the conditional distribution $P[Q_{t+1} | Q_t, \ldots, Q_1]$. Here, the first-order Markov assumption is defined by

$$P[Q_{t+1} | Q_t, \ldots, Q_1] := P[Q_{t+1} | Q_t] \tag{2.2}$$

to model statistical dependencies only between the next random variable Q_{t+1} and its direct predecessor Q_t. This assumption is frequently used in different applications including the analysis of DNA sequences (Durbin et al. (1998)), or considering the analysis of wind direction, social behavior, and financial time series data (Berchtold and Raftery (2002)). The first-order Markov assumption in (2.2) can be defined as being independent of the specific time step t or as being time-dependent. Both cases are considered subsequently.

2.1.1 Homogeneous First-Order Markov Model

A homogeneous first-order *MM* models statistical dependencies between a next random variable Q_{t+1} and its direct predecessor random variable Q_t independent of the time step t. That is, the attribute 'homogeneous' defines that the probability of observing $Q_{t+1} = j$ given that $Q_t = i$ is identical for each time step t. Based on this, the homogeneous first-order *MM* $\lambda = (\vec{\pi}, A)$ is defined by the following parameters.

1. The initial state distribution $\vec{\pi} := (\pi_{S_1}, \ldots, \pi_{S_N})$ defines for each state $i \in S$ the probability $\pi_i := P[Q_1 = i]$ of starting in this state at time step $t = 1$. Two stochastic constraints must be fulfilled by $\vec{\pi}$.

    a) $\forall i \in S : \pi_i \in [0, 1]$

2. Markov Models

b) $\sum_{i \in S} \pi_i = 1$

2. The transition matrix $A = (a_{ij})$ defines for each current state $i \in S$ and each next state $j \in S$ the transition probability $a_{ij} := P[Q_{t+1} = j \mid Q_t = i]$ for the transition from i to j at all time steps t. Each row $i \in S$ of A has to fulfill two stochastic constraints.

   a) $\forall j \in S : a_{ij} \in [0, 1]$

   b) $\sum_{j \in S} a_{ij} = 1$

For a state sequence $\vec{q} := (q_1, \ldots, q_T)$ with $q_t \in S$ for each time step $1 \leq t \leq T$, the likelihood of $\vec{q}$ under the homogeneous first-order MM λ is given by

$$P[\vec{Q} = \vec{q} \mid \lambda] = P[Q_1 = q_1 \mid \lambda] \cdot \prod_{t=1}^{T-1} P[Q_{t+1} = q_{t+1} \mid Q_t = q_t, \lambda]$$

$$= \pi_{q_1} \cdot \prod_{t=1}^{T-1} a_{q_t q_{t+1}}.$$

This formula is obtained based on the joint distribution in (2.1) that is modified by integrating the first-order Markov assumption in (2.2) in consideration that λ represents a homogeneous MM. The statistical dependencies modeled by a homogeneous first-order MM can be represented by a first-order Markov chain shown in Fig. 2.1. A state-transition diagram like that shown in Fig. 2.2 can be used to visualize the initial state distribution and the transition matrix of a specific homogeneous first-order MM.

Figure 2.1: Markov chain $\vec{Q} := (Q_1, \ldots, Q_T)$ represented by a homogeneous first-order MM. Each data point modeled by the random variable Q_{t+1} is assumed to be depending on its direct predecessor data point modeled by the random variable Q_t. This assumption corresponds to a graph with links between consecutive random variables.

2.1.2 Inhomogeneous First-Order Markov Model

The inhomogeneous first-order MM extends the homogeneous first-order MM by realizing time-dependent state-transitions. This means, the transition from the current state modeled by the random variable Q_t to the next state modeled by the random variable

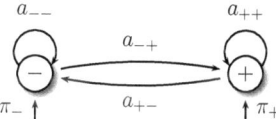

Figure 2.2: State-transition diagram of a two-state homogeneous first-order *MM* with states $S := \{-, +\}$ represented by labeled circles. The initial state probability of each state $i \in S$ is given by the corresponding arrow labeled with π_i. The transition probability of a transition from a current state $i \in S$ to a next state $j \in S$ is represented by the arrow labeled with a_{ij}.

Q_{t+1} is explicitly depending on the time step t at which the transition is done. The drawback of this is the increase in the number of transition parameters that are required to model state sequences. Here, to overcome this, the finite set $\mathcal{C} := \{1, \ldots, C\}$ of transition classes is introduced to reduce the number of transition parameters. Based on this, for each transition class $c \in \mathcal{C}$ a corresponding transition matrix A_c is defined in analogy to the transition matrix of the homogeneous first-order *MM*.

- The transition matrix $A_c = (a_{ij}(c))$ defines for each current state $i \in S$ and each next state $j \in S$ the transition probability $a_{ij}(c) := P[Q_{t+1} = j \,|\, Q_t = i, c]$ for the transition from i to j at time step t using the transition class c. Each row $i \in S$ of A_c has to fulfill two stochastic constraints.

    1. $\forall j \in S : a_{ij}(c) \in [0, 1]$
    2. $\sum_{j \in S} a_{ij}(c) = 1$

All transition matrices of the inhomogeneous first-order *MM* are represented by the set of transition matrices $A := \{A_1, \ldots, A_C\}$. Time-dependent transitions between states are realized by defining a transition class sequence $\vec{c} := (c_1, \ldots, c_{T-1})$ with $c_t \in \mathcal{C}$ for each time step $1 \leq t < T$. Each c_t specifies that the transition matrix A_{c_t} has to be used at time step t for the transition from the current state modeled by the random variable Q_t to the next state modeled by the random variable Q_{t+1}. That means, for a state sequence $\vec{q} = (q_1, \ldots, q_T)$ the probability for the transition from q_t to q_{t+1} is given by $a_{q_t q_{t+1}}(c_t)$. The likelihood of a state sequence $\vec{q}$ with respect to a transition class

2. Markov Models

sequence $\vec{c}$ is given by

$$P[\vec{Q} = \vec{q} \mid \vec{c}, \lambda] = P[Q_1 = q_1 \mid \lambda] \cdot \prod_{t=1}^{T-1} P[Q_{t+1} = q_{t+1} \mid Q_t = q_t, c_t, \lambda]$$

$$= \pi_{q_1} \cdot \prod_{t=1}^{T-1} a_{q_t q_{t+1}}(c_t)$$

under the inhomogeneous first-order *MM* λ. The formula of the likelihood is obtained in analogy to that of the homogeneous first-order *MM* by additionally integrating the transition class sequence for realizing time-dependent state-transitions. For $C = 1$ transition class, the inhomogeneous first-order *MM* simplifies to the homogeneous first-order *MM*.

To illustrate the time-dependent state-transition process, the state-transition diagram of an inhomogeneous first-order *MM* with two states and two transition classes is shown in Fig. 2.3. A corresponding Markov chain of length 6 that is realized by this model for a given transition class sequence $\vec{c} = (1, 2, 2, 1, 2)$ is exemplarily shown in Fig. 2.4. In practical applications, the usage of transition classes provides the opportunity to integrate additional information into the state-transition process. This has initially been used by Knab et al. (2003) to improve the clustering of financial time-series data. In computational biology, transition classes can be used to integrate chromosomal distances and orientations of adjacent genes on a chromosome (Seifert et al. (2009b)).

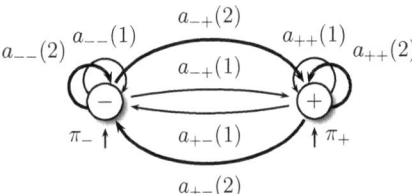

Figure 2.3: State-transition diagram for an inhomogeneous first-order *MM* with two states $S := \{-, +\}$ and two transition classes $C := \{1, 2\}$. The states are represented by labeled circles. The initial state probability of each state $i \in S$ is given by the corresponding arrow labeled with π_i. The transition probability of a transition from a current state $i \in S$ to a next state $j \in S$ in transition class 1 is represented by the thin arrow labeled with $a_{ij}(1)$, and that of transition class 2 is represented by the thick arrow labeled with $a_{ij}(2)$.

Figure 2.4: A Markov chain $\vec{Q} := (Q_1, \ldots, Q_6)$ of length 6 realized by the *MM* in Fig. 2.3 for a given transition class sequence $\vec{c} = (1, 2, 2, 1, 2)$. Each data point modeled by the random variable Q_{t+1} is assumed to be depending on its direct predecessor data point modeled by the random variable Q_t. Transitions between Q_t and Q_{t+1} that are defined to be done via transition class 1 are represented by thin links, and those that are defined to use the transition class 2 are represented by thick links.

2.2 Higher-Order Markov Models

The first-order *MM* only represents statistical dependencies between the next state modeled by Q_{t+1} and its direct predecessor state modeled by Q_t. Since several of the most recent consecutive predecessor states might provide useful information for the next state, the first-order Markov assumption in (2.2) could be too restrictive. To overcome this, the L-th order Markov assumption defined by

$$P[Q_{t+1} | Q_t, \ldots, Q_1] := \begin{cases} P[Q_{t+1} | \vec{Q}_{1 \ldots t}], & 1 \leq t < L \\ P[Q_{t+1} | \vec{Q}_{t-L+1 \ldots t}], & t \geq L \end{cases} \quad (2.3)$$

is used to represent statistical dependencies between Q_{t+1} and its most recent predecessors $\vec{Q}_{max(1,t-L+1) \ldots t} := (Q_{max(1,t-L+1)}, \ldots, Q_t)$. This assumption is modeled by a higher-order *MM* of order L (Berchtold and Raftery (2002)). This *MM* has a memory to store the most recent predecessor states for modeling statistical dependencies between these states and the next state. During the first time steps $1 \leq t \leq L$, the memory is filled up by adding the current state $q_t \in S$ modeled by Q_t until the memory represents the state context $\vec{q}_{1 \ldots L} = (q_1, \ldots, q_L)$. At each later time step $t > L$, the memory changes to $\vec{q}_{t-L+1 \ldots t} = (q_{t-L+1}, \ldots, q_t)$ by deleting the oldest state q_{t-L} and by adding the new current state q_t. In analogy to the first-order Markov assumption in (2.2), the L-th order Markov assumption in (2.3) can be defined as being independent of the time step t or as being time-dependent. This is considered subsequently for the homogeneous higher-order *MM* for which transitions between states are independent of the current time step, and for the inhomogeneous higher-order *MM* that realizes time-dependent state-transitions.

2. Markov Models

2.2.1 Homogeneous Higher-Order Markov Model

The homogeneous higher-order *MM* of order L has a memory to model the statistical dependencies expressed by the L-th order Markov assumption in (2.3). To specify the states stored in this memory, the definition of the set of state contexts

$$S^l := \{(i_1, \ldots, i_l) : i_1 \in S, \ldots, i_l \in S\}$$

of length $l \in \mathbb{N}$ is required. Each state context $i = (i_1, \ldots, i_l)$ is contained in S^l. Based on this, the transition matrix A of the homogeneous first-order *MM* is subsequently extended to that of the homogeneous higher-order *MM* of order L.

- The transition matrix $A = (a_{ij})$ defines for each state context $i = (i_1, \ldots, i_l) \in S^l$ of length $1 \leq l \leq L$ and for each next state $j \in S$ the transition probability

$$a_{ij} := \begin{cases} P[Q_{t+1} = j \,|\, \vec{Q}_{1\ldots t} = i], & 1 \leq t < L \\ P[Q_{t+1} = j \,|\, \vec{Q}_{t-L+1\ldots t} = i], & t \geq L \end{cases}$$

for a transition from the current state i_l to the next state j at all time steps t with respect to the predecessor states $(i_1, \ldots, i_{l-1})$ of the current state. Again, each row i of A has to fulfill two stochastic constraints.

a) $\forall j \in S : a_{ij} \in [0, 1]$

b) $\sum_{j \in S} a_{ij} = 1$

The number of transition parameters increases from N^2 for the first-order *MM* to $\sum_{l=1}^{L} N^{l+1} = N((1 - N^{L+1})/(1 - N) - 1)$ for the homogeneous *MM* of order L. In contrast to this, the initial state distribution $\vec{\pi}$ defined for the first-order *MM* can be used without any adaptations. For order $L = 1$, the homogeneous *MM* of order L represents the homogeneous first-order *MM* as a special case. In analogy to the homogeneous first-order *MM*, the likelihood of a state sequence $\vec{q} = (q_1, \ldots, q_T)$ is represented by

$$P[\vec{Q} = \vec{q} \,|\, \lambda] = P[Q_1 = q_1 \,|\, \lambda] \cdot \prod_{t=1}^{L-1} P[Q_{t+1} = q_{t+1} \,|\, \vec{Q}_{1\ldots t} = \vec{q}_{1\ldots t}, \lambda]$$

$$\cdot \prod_{t=L}^{T-1} P[Q_{t+1} = q_{t+1} \,|\, \vec{Q}_{t-L+1\ldots t} = \vec{q}_{t-L+1\ldots t}, \lambda]$$

$$= \pi_{q_1} \cdot \prod_{t=1}^{L-1} a_{\vec{q}_{1\ldots t} q_{t+1}} \cdot \prod_{t=L}^{T-1} a_{\vec{q}_{t-L+1\ldots t} q_{t+1}}$$

under the homogeneous *MM* λ of order L. This formula results from the joint distribution in (2.1) that is modified with respect to the L-th order Markov assumption in (2.3). The statistical dependencies modeled by a homogeneous *MM* of order two are exemplarily shown in Fig. 2.5. For further reading to homogeneous higher-order *MM*s one can consider the review by Berchtold and Raftery (2002) or the textbook by Bishop (2006). In computational biology, higher-order *MM*s are frequently used as background models for the analysis of DNA sequences (Durbin et al. (1998)).

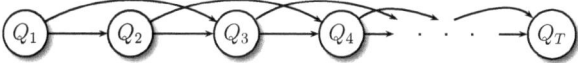

Figure 2.5: Markov chain $\vec{Q} := (Q_1, \ldots, Q_T)$ represented by a *MM* of order two. Each data point modeled by the random variable Q_{t+1} is assumed to be depending on its two direct predecessor data points modeled by the random variables Q_t and Q_{t-1}. This assumption corresponds to a graph with links between each Q_{t+1} and its two predecessors Q_t and Q_{t-1}.

2.2.2 Inhomogeneous Higher-Order Markov Model

The inhomogeneous *MM* of order L extends the homogeneous *MM* of order L by realizing time-dependent state-transitions. Now, the transition from the current state modeled by Q_t to the next state modeled by Q_{t+1} under consideration of the most recent predecessor states modeled by $\vec{Q}_{\max(1,t-L+1)\ldots t}$ is explicitly depending on the time step t at which the transition is done. Like introduced for the inhomogeneous first-order *MM*, the finite set of $\mathcal{C}$ of transition classes is used to reduce the number of transition parameters. Based on this, a transition matrix A_c is defined for each transition class $c \in \mathcal{C}$ in analogy to the transition matrix of the homogeneous higher-order *MM*.

- The transition matrix $A_c = (a_{ij}(c))$ defines for each state context $i = (i_1, \ldots, i_l) \in S^l$ of length $1 \leq l \leq L$ and for each next state $j \in S$ the transition probability

$$a_{ij}(c) := \begin{cases} P[Q_{t+1} = j \mid \vec{Q}_{1\ldots t} = i, c], & 1 \leq t < L \\ P[Q_{t+1} = j \mid \vec{Q}_{t-L+1\ldots t} = i, c], & t \geq L \end{cases}$$

for a transition from the current state i_l to the next state j at time step t using the transition class c with respect to the predecessor states $(i_1, \ldots, i_{l-1})$ of the current state. Each row i of A_c has to fulfill two stochastic constraints.

a) $\forall j \in S : a_{ij}(c) \in [0, 1]$

2. Markov Models

b) $\sum_{j \in S} a_{ij}(c) = 1$

Time-dependent transitions are again realized by the transition class sequence $\vec{c} = (c_1, \ldots, c_{T-1})$. This sequence specifies which transition matrix A_{c_t} has to be used for the transition at a specific time step t. Under consideration of this, the likelihood of a state sequence $\vec{q} = (q_1, \ldots, q_T)$ is given by

$$P[\vec{Q} = \vec{q} \,|\, \vec{c}, \lambda] = \pi_{q_1} \cdot \prod_{t=1}^{L-1} a_{\vec{q}_{1 \ldots t} q_{t+1}}(c_t) \cdot \prod_{t=L}^{T-1} a_{\vec{q}_{t-L+1 \ldots t} q_{t+1}}(c_t)$$

under the inhomogeneous MM λ of order L with respect to $\vec{c}$. This formula extends the likelihood function given for the homogeneous MM of order L by integrating the transition class sequence to enable time-dependent transitions. Since the inhomogeneous MM of order L uses C transition classes, the number of transition parameters increases by a factor of C in comparison to the homogeneous MM of order L. The inhomogeneous MM of order L with C transition classes reduces to the homogeneous first-order MM for $L = 1$ and $C = 1$, simplifies to the homogeneous higher-order MM for $L > 1$ and $C = 1$, and does represent the inhomogeneous first-order MM for $L = 1$ and $C > 1$. Thus, the inhomogeneous higher-order MM introduced here provides a good basis to extend the theory behind homogeneous first-order HMM in the three following chapters.

3 Hidden Markov Models

The goal of this chapter is to develop the theory of the inhomogeneous Higher-order Hidden Markov Model of order L with C transition classes ($HHMM(L,C)$) by extending the standard theory of the homogeneous first-order Hidden Markov Model (*HMM*). The basis of the homogeneous *HMM* is the homogeneous first-order *MM* that has been introduced in the previous chapter. This *MM* is extended to the *HMM* by adding a state-specific emission process that enables the processing of sequential data that is typically referred to as emission sequence in the context of *HMM*s. In analogy to this, the inhomogeneous $HHMM(L,C)$ is based on the inhomogeneous higher-order *MM*. Thus, the state-transition process of the $HHMM(L,C)$ accounts for the predecessor states by introducing a memory of size L, and additional knowledge is integrated into this process by using C transition classes. These extensions require the adaptation of the standard algorithms of the homogeneous *HMM* to algorithms for the inhomogeneous $HHMM(L,C)$. For that reason, the Forward algorithm, the Backward algorithm, the State-Posterior algorithm, the Viterbi algorithm, and the Baum-Welch algorithm are extended. Good introductions to these standard algorithms of the homogeneous *HMM* are given in the review by Rabiner (1989) and in the textbooks by Durbin et al. (1998) and Bishop (2006). The extension of these standard algorithms are used to solve the following four basic problems that frequently occur in *HMM*-based data analysis.

1. *Likelihood Problem*: How can one compute the likelihood of an emission sequence $\vec{o}(k)$ under an inhomogeneous $HHMM(L,C)$ with respect to a given transition class sequence $\vec{c}(k)$?

2. *Optimal State Sequence Problem*: How can one choose a state sequence $\vec{q}$ that is optimal for a given emission sequence $\vec{o}(k)$ under an inhomogeneous $HHMM(L,C)$ with respect to a given transition class sequence $\vec{c}(k)$?

3. *Maximum Likelihood Problem*: How can one adjust the parameters of an inhomogeneous $HHMM(L,C)$ to maximize the likelihood of emission sequences

3. Hidden Markov Models

$\vec{o}(1), \ldots, \vec{o}(K)$ with respect to their transition class sequences $\vec{c}(1), \ldots, \vec{c}(K)$ under this model?

4. *Maximum A Posteriori Problem*: How can one adjust the parameters of an inhomogeneous *HHMM*(L, C) to maximize the posterior of this model in consideration of emission sequences $\vec{o}(1), \ldots, \vec{o}(K)$ and their transition class sequences $\vec{c}(1), \ldots, \vec{c}(K)$?

Besides the extension of the standard algorithms and the development of solutions for the four basic problems, the inhomogeneous *HHMM*(L, C) also provides the basics for two specific model extensions, the parsimonious higher-order *HMM* and the *HMM* with scaled transition matrices, which are introduced in the two following chapters. Additionally, the development of extended standard algorithms also establishes the basis for the application of these models in different research fields. Most of the published articles on *HMM*s belong to the field of speech recognition. Two of the best reviews for getting an overview in this research field have been published by Rabiner (1989) and by Ephraim and Merhav (2002). Again in the field of speech recognition, first applications of the homogeneous second-order *HHMM*(2) were published in a series of articles by Kriouile et al. (1990), Mari and Haton (1994), Mari et al. (1996), and by Mari et al. (1997). More recent publications in this field, like de Villiers and du Preez (2001) and Lee and Lee (2006) also focus on the homogeneous *HHMM*(L). Other application fields of the homogeneous *HHMM*(L) are image segmentation (Derrode et al. (2004); Mari and Le Ber (2006); Benyoussef et al. (2008)), robotics (Aycard et al. (2004)), and modeling of DNA sequences (Ching et al. (2003)). All these different application fields define an excellent starting point to develop the theory behind the inhomogeneous *HHMM*(L, C).

Goals of this Chapter

1. The extension of Markov Models to Hidden Markov Models is motivated briefly by considering the basics of the homogeneous first-order *HMM*.

2. The definition of the inhomogeneous *HHMM*(L, C) is given.

3. The Forward algorithm and the Backward algorithm are extended to provide a solution to the *Likelihood Problem*.

4. The State-Posterior algorithm and the Viterbi algorithm are extended to provide two solutions of the *Optimal State Sequence Problem*.

5. Based on the extension of the Baum-Welch algorithm a solution to the *Maximum Likelihood Problem* is developed.

6. A prior for the *HHMM*(L, C) is introduced to enable the integration of prior knowledge into the training of the model.

7. Under consideration of the prior, a Bayesian version of the Baum-Welch algorithm is developed to provide a solution of the *Maximum A Posteriori Problem*.

3.1 Homogeneous First-Order Hidden Markov Models

The basis of modeling sequential data by an *HMM* with continuous emissions is given by the joint density

$$P[\vec{O},\vec{Q}] = P[Q_1] \cdot \prod_{t=1}^{T-1} P[Q_{t+1}|Q_t] \cdot \prod_{t=1}^{T} P[O_t|Q_t] \tag{3.1}$$

for a sequence of continuous random variables $\vec{O} := (O_1, \ldots, O_T)$ and a sequence of discrete random variables $\vec{Q} := (Q_1, \ldots, Q_T)$. In the following, each random variable O_t is defined to model an emission over the set of real numbers $\mathbb{R}$ in dependency of the random variable Q_t that is defined to model a state by taking values in the finite set of states $S := \{S_1, \ldots, S_N\}$. The joint density in (3.1) is known as the state space model that forms the basis of the *HMM* (Bishop (2006)). The statistical dependencies represented by this model are visualized in Fig. 3.1. Thus, the joint density in (3.1) extends the state-transition process of the homogeneous first-order *MM* shown in Fig. 2.1 by an additional stochastic emission process. For the *HMM*, this emission process is realized by the time-independent emission density $P[O_t|Q_t]$. That means, the probability density $b_i(o) := P[O_t = o | Q_t = i]$ of observing the emission o under state i is identical for all time steps t. Subsequently, the focus is on *HMM*s with Gaussian emission densities. For that reason, each state $i \in S$ is characterized by its corresponding Gaussian emission density

$$b_i(o) := \frac{1}{\sqrt{2\pi}\sigma_i} \exp\left(-\frac{(o-\mu_i)^2}{2\sigma_i^2}\right) \tag{3.2}$$

to represent the probability density of an emission $o \in \mathbb{R}$ under state i in consideration of the state-specific mean $\mu_i \in \mathbb{R}$ and the state-specific standard deviation $\sigma_i \in \mathbb{R}^+$.

3. Hidden Markov Models

Due to the precense of the emission process, the state-transition process of the *HMM* is hidden. That means, an emission o_t modeled by O_t cannot be assigned uniquely to a specific state q_t modeled by Q_t. Instead, the emission o_t is assigned to each state q_t of the *HMM* according to the emission density $b_{q_t}(o_t)$ specified in (3.2). This property is very useful for the analysis of an emission sequence $\vec{o} := (o_1, \ldots, o_T)$, because it enables the assignment of a characteristic state q_t to each individual emission o_t for interpreting the whole emission sequence under an *HMM*.

Since the homogeneous first-order *HMM* extends the homogeneous first-order *MM* by the state-specific emission process in (3.2), the joint density in (3.1) for an emission sequence $\vec{o} = (o_1, \ldots, o_T)$ and a state sequence $\vec{q} = (q_1, \ldots, q_T)$ is given by

$$P[\vec{O} = \vec{o}, \vec{Q} = \vec{q} \mid \lambda] = \pi_{q_1} \cdot \prod_{t=1}^{T-1} a_{q_t q_{t+1}} \cdot \prod_{t=1}^{T} b_{q_t}(o_t)$$

under the homogeneous first-order *HMM* λ. This joint density is generally referred to as the complete-data likelihood since the specific hidden state sequence $\vec{q}$ that emits the corresponding emission sequence $\vec{o}$ is assumed to be known. For a more detailed introduction to homogeneous first-order *HMM*s, the review by Rabiner (1989) and the textbooks by Durbin et al. (1998) and Bishop (2006) can be considered. Here, in this thesis, the focus is on the extension of the homogeneous first-order *HMM* based on the different *MM*s introduced in previous chapter.

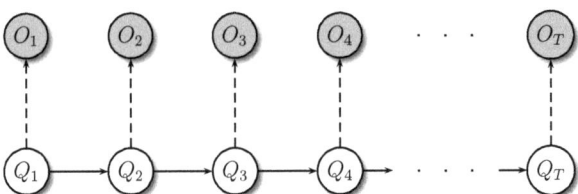

Figure 3.1: State space model that underlies a homogeneous first-order *HMM*. The Markov chain $\vec{Q} = (Q_1, \ldots, Q_T)$ defined by the homogeneous first-order *MM* is the internal state-transition system of the *HMM*. Each transition to a next state modeled by the random variable Q_{t+1} is depending on the direct predecessor state modeled by the random variable Q_t. The Markov chain $\vec{Q}$ is hidden. Only the emission modeled by the random variable O_t that is made in dependency of the corresponding state modeled by Q_t is visible.

3.2 Inhomogeneous Higher-Order Hidden Markov Model

The internal system that realizes the state-transition process of the *HHMM*(L, C) is the inhomogeneous higher-order *MM* specified in Sec. 2.2.2. This *MM* is extended by an additional stochastic emission process like described in the previous section for the homogeneous first-order *HMM*. Thus, the inhomogeneous *HHMM*(L, C) of order L with C different transition classes is defined by $\lambda = (\vec{\pi}, A, B)$ in consideration of the following parameters.

1. The initial state distribution $\vec{\pi} := (\pi_{S_1}, \ldots, \pi_{S_N})$ defines for each state $i \in S$ the probability $\pi_i := P[Q_1 = i]$ of starting in this state at time step $t = 1$. Two stochastic constraints must be fulfilled by $\vec{\pi}$.

    a) $\forall i \in S : \pi_i \in [0, 1]$
    b) $\sum_{i \in S} \pi_i = 1$

2. The set $A := \{A_1, \ldots, A_C\}$ defines the C transition class specific transition matrices. For each transition class $c \in \mathcal{C}$, the transition matrix $A_c = (a_{ij}(c))$ defines for each state context $i = (i_1, \ldots, i_l) \in S^l$ of all lengths $1 \leq l \leq L$ and for each next state $j \in S$ the transition probability

$$a_{ij}(c) := \begin{cases} P[Q_{t+1} = j \,|\, \vec{Q}_{1\ldots t} = i, c], & 1 \leq t < L \\ P[Q_{t+1} = j \,|\, \vec{Q}_{t-L+1\ldots t} = i, c], & t \geq L \end{cases}$$

    for a transition from the current state i_l to the next state j at time step t using the transition class c with respect to the memory $(i_1, \ldots, i_{l-1})$ of the current state. Again, each row i of A_c has to fulfill two stochastic constraints.

    a) $\forall j \in S : a_{ij}(c) \in [0, 1]$
    b) $\sum_{j \in S} a_{ij}(c) = 1$

3. The matrix $B := (\mu_i, \sigma_i)$ defines the state-specific mean $\mu_i \in \mathbb{R}$ and the state-specific standard deviation $\sigma_i \in \mathbb{R}^+$ for the Gaussian emission density of each state $i \in S$. The time-independent probability density $b_i(o) := P[O_t = o \,|\, Q_t = i]$ for emitting an emission $o \in \mathbb{R}$ by the Gaussian emission density of state i is defined in (3.2).

The *HHMM*(L, C) specified here represents a very general class of models. With respect to the notation scheme in Tab. 3.1, the inhomogeneous *HHMM*(L, C) reduces to the homogeneous first-order *HMM* for $L = 1$ and $C = 1$, represents the inhomogeneous first-order *HMM(C)* for $L = 1$ and $C > 1$, and defines the homogeneous higher-order *HHMM(L)* for $L > 1$ and $C = 1$. The *HMM(C)* has been introduced by Knab et al. (2003) for the analysis of financial time-series data. The basis of this model is an inhomogeneous first-order *MM*. The *HHMM(L)* is based on the homogeneous higher-order *MM*.

Notation	C	L	Hidden Markov Model
HMM	1	1	homogeneous first-order
HMM(C)	>1	1	inhomogeneous first-order
HHMM(L)	1	>1	homogeneous higher-order
HHMM(L,C)	>1	>1	inhomogeneous higher-order

Table 3.1: Basic notation scheme of Hidden Markov Models that can be represented by the *HHMM(L,C)*. The order is given by L and the number of transition classes is defined by C.

3.3 Solving the Likelihood Problem

One strategy to compute the likelihood $P[\vec{o}(k) \,|\, \vec{c}(k), \lambda]$ of an emission sequence $\vec{o}(k) = (o_1(k), \ldots, o_{T_k}(k))$ given a transition class sequence $\vec{c}(k) = (c_1(k), \ldots, c_{T_k-1}(k))$ and an inhomogeneous *HHMM(L,C)* λ is to marginalize the complete-data likelihood $P[\vec{o}(k), \vec{q} \,|\, \vec{c}(k), \lambda]$ over each individual state sequence $\vec{q} \in S^{T_k}$. The fundamental drawback of this is the exponential increase of the number of possible state sequences for increasing T_k that have to be considered separately. This increase is illustrated in Fig. 3.2. To overcome this, the Forward algorithm and the Backward algorithm are introduced to efficiently compute the likelihood by marginalization based on a dynamic programming approach. Both algorithms are standard computational tools for the homogeneous *HMM* (Rabiner (1989); Durbin et al. (1998); Bishop (2006)). Here, both algorithms are extended for the inhomogeneous *HHMM(L,C)*. Based on that, the likelihood can be computed, and basics for solving the three other basic problems are provided. Additionally, the Forward-Backward procedure for computing the likelihood is deduced from both algorithms.

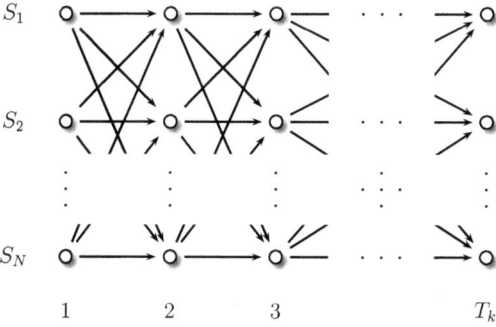

Figure 3.2: State sequence space representing each individual state sequence that models a given emission sequence $\vec{o}(k)$ of length T_k. The number of possible state sequences grows exponentially for increasing T_k with growth rate N given by the number of hidden states of the underlying inhomogeneous *HHMM*(L, C).

3.3.1 Forward Algorithm

The Forward-Algorithm is an efficient dynamic programming approach for computing the likelihood in three steps. This algorithm is based on the definition of the Forward-Variable

$$\alpha_t^k(i) := \begin{cases} P[\vec{o}_{1\ldots t}(k), \vec{Q}_{1\ldots t} = i \,|\, \vec{c}_{1\ldots t-1}(k), \lambda] & , 1 \leq t < L \\ P[\vec{o}_{1\ldots t}(k), \vec{Q}_{t-L+1\ldots t} = i \,|\, \vec{c}_{1\ldots t-1}(k), \lambda] & , L \leq t \leq T_k \end{cases} \quad (3.3)$$

which represents, in consideration of time step t, the joint probability density of emitting the partial emission sequence $\vec{o}_{1\ldots t}(k)$ and having the state context $i \in S^{\min(t,L)}$ given the partial transition class sequence $\vec{c}_{1\ldots t-1}(k)$ and the inhomogeneous $HHMM(L,C)$ λ. The likelihood of an emission sequence $\vec{o}(k)$ of length $T_k \geq L$ is computed by the following algorithm based on the iterative computation of the Forward-Variables.

- *Initialization*

$$\forall i = (i_1) \in S^1: \quad \alpha_1^k(i) = \pi_{i_1} \cdot b_{i_1}(o_1(k))$$

- *Induction*

$$\forall 1 \leq t < L \text{ and } \forall i = (i_1, \ldots, i_{t+1}) \in S^{t+1} \text{ with } j = (i_1, \ldots, i_t)$$

$$\alpha_{t+1}^k(i) = \alpha_t^k(j) \cdot a_{j i_{t+1}}(c_t(k)) \cdot b_{i_{t+1}}(o_{t+1}(k))$$

$$\forall L \leq t < T_k \text{ and } \forall i = (i_1, \ldots, i_L) \in S^L \text{ with } j(i_0) = (i_0, i_1, \ldots, i_{L-1})$$

$$\alpha_{t+1}^k(i) = \sum_{i_0 \in S} \alpha_t^k(j(i_0)) \cdot a_{j(i_0) i_L}(c_t(k)) \cdot b_{i_L}(o_{t+1}(k))$$

- *Termination*

$$P[\vec{o}(k) \,|\, \vec{c}(k), \lambda] = \sum_{i \in S^L} \alpha_{T_k}^k(i)$$

In the initialization step each state-specific Forward-Variable for state $i_1 \in S$ is computed by multiplying the initial state probability π_{i_1} with the corresponding emission density $b_{i_1}(o_1(k))$ of the first emission $o_1(k)$ of $\vec{o}(k)$. The induction step first computes each Forward-Variable $\alpha_{t+1}^k(i)$ for time step $1 \leq t < L$ as the joint density of passing through the states $\vec{q}_{1\ldots t+1} = i \in S^{t+1}$ for emitting the corresponding partial emission sequence $\vec{o}_{1\ldots t+1}(k)$ of $\vec{o}(k)$. Then, in the second part of the induction step, each Forward-Variable $\alpha_{t+1}^k(i)$ for time step $L \leq t < T_k$ is computed as the joint density of the partial emission sequence $\vec{o}_{1...t+1}(k)$ and the state context $\vec{q}_{t-L+2...t+1} = (i_1, \ldots, i_L) \in S^L$ by marginalizing over all N states $q_{t-L+1} = i_0 \in S$ that occurred $L+1$ time steps before the current state $q_{t+1} = i_L$ is reached for emitting the corresponding emission $o_{t+1}(k)$ of $\vec{o}(k)$. The computational scheme that underlies this part of the induction step is illustrated in Fig. 3.3. Finally, the termination step gives the desired likelihood of an emission sequence $\vec{o}(k)$ under an inhomogeneous HHMM(L, C) λ with respect to the given transition class sequence $\vec{c}(k)$ by marginalizing over all terminal Forward-Variables $\alpha^k_{T_k}(i)$. According to that, the Forward algorithm has a total run-time of $O\left((T_k - L)N^{L+1}\right)$. This follows from the second part of the induction step that considers $T_k - L$ time steps in which the computation of one of the N^L Forward-Variables for one time step requires $O(N)$ operations. Basic introductions to the Forward algorithm for a homogeneous HMM are given by Rabiner (1989), Durbin et al. (1998), and Bishop (2006). An extension of the Forward algorithm to a homogeneous HHMM(L) is given by Ching et al. (2003).

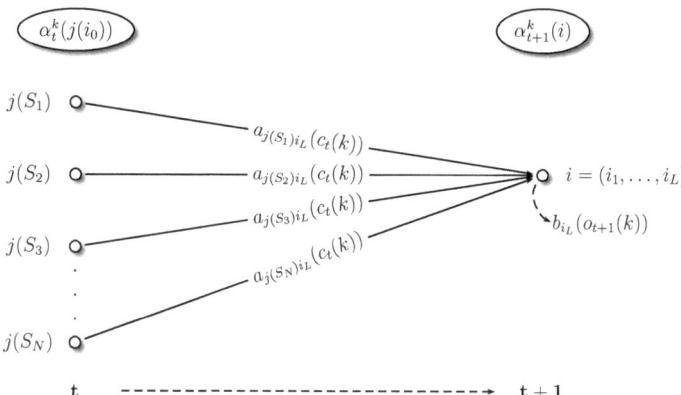

Figure 3.3: Computational scheme of the Forward-Variable $\alpha^k_{t+1}(i)$ of state context $i = (i_1, \ldots, i_L) \in S^L$ during the second part of the induction step. Each Forward-Variable $\alpha^k_t(j(i_0))$ of state context $j(i_0) = (i_0, i_1, \ldots, i_{L-1})$ with $i_0 \in S$ is considered by transforming $j(i_0)$ to i. This is done by the transition from the current state i_{L-1} to i_L using the transition probability $a_{j(i_0)i_L}(c_t(k))$. After this transition the emission $o_{t+1}(k)$ is made by state i_L using its emission density $b_{i_L}(o_{t+1}(k))$.

3.3.2 Backward Algorithm

The Backward-Algorithm is another efficient dynamic programming approach for computing the likelihood in three steps. The basis of this algorithm is the Backward-Variable

$$\beta_t^k(i) := \begin{cases} P[\vec{o}_{t+1\ldots T_k}(k) \,|\, \vec{Q}_{t-L+1\ldots t} = i, \vec{c}_{t\ldots T_k-1}(k), \lambda] & , L \leq t \leq T_k \\ P[\vec{o}_{t+1\ldots T_k}(k) \,|\, \vec{Q}_{1\ldots t} = i, \vec{c}_{t\ldots T_k-1}(k), \lambda] & , 1 \leq t < L \end{cases} \quad (3.4)$$

which defines, in consideration of time step t, the probability density of emitting the partial emission sequence $\vec{o}_{t+1\ldots T_k}(k)$ given the state context $i \in S^{min(t,L)}$, the partial transition class sequence $\vec{c}_{t\ldots T_k-1}(k)$, and the inhomogeneous $HHMM(L,C)$ λ. The likelihood of an emission sequence $\vec{o}(k)$ of length $T_k \geq L$ is computed by the following algorithm based on the iterative computation of the Backward-Variables.

- *Initialization*

  $\forall i \in S^L: \quad \beta_{T_k}^k(i) = 1$

- *Induction*

  $\forall T_k \geq t > L$ and $\forall i = (i_1, \ldots, i_L) \in S^L$ with $j(i_{L+1}) = (i_2, \ldots, i_L, i_{L+1})$

  $$\beta_{t-1}^k(i) = \sum_{i_{L+1} \in S} \beta_t^k(j(i_{L+1})) \cdot a_{ii_{L+1}}(c_{t-1}(k)) \cdot b_{i_{L+1}}(o_t(k))$$

  $\forall L \geq t > 1$ and $\forall i = (i_1, \ldots, i_{t-1}) \in S^{t-1}$ with $j(i_t) = (i_1, \ldots, i_{t-1}, i_t)$

  $$\beta_{t-1}^k(i) = \sum_{i_t \in S} \beta_t^k(j(i_t)) \cdot a_{ii_t}(c_{t-1}(k)) \cdot b_{i_t}(o_t(k))$$

- *Termination*

  $$P[\vec{o}(k) \,|\, \vec{c}(k), \lambda] = \sum_{i=(i_1) \in S^1} \pi_{i_1} \cdot b_{i_1}(o_1(k)) \cdot \beta_1^k(i)$$

The initialization step sets all Backward-Variables $\beta_{T_k}^k(i)$ to one to provide the basics for the induction step. The first part of the induction step computes all Backward-Variables $\beta_{t-1}^k(i)$ of time steps $T_k \geq t > L$ for emitting the partial emission sequence $\vec{o}_{t\ldots T_k}(k)$ given the state context $\vec{q}_{t-L\ldots t-1} = i \in S^L$ with $i = (i_1, \ldots, i_L)$ and the partial transition class sequence $\vec{c}_{t-1\ldots T_k-1}(k)$. All possible next states $q_t = i_{L+1} \in S$ have to be considered. First, this is done by the transition probability $a_{ii_{L+1}}(c_{t-1}(k))$, which accounts for the transition from the current state $q_{t-1} = i_L$ to the next state $q_t = i_{L+1}$ in consideration of the memory $\vec{q}_{t-L...t-2} = (i_1, \ldots, i_{L-1})$ of the current state. Next, the emission $o_t(k)$ done by the next state $q_t = i_{L+1}$ is integrated by $b_{i_{L+1}}(o_t(k))$. The remaining partial emission sequence $\vec{o}_{t+1...T_k}(k)$ is already represented by the Backward-Variable $\beta_t^k(j(i_{L+1}))$. This first part of the induction step is illustrated in the computational scheme shown in Fig. 3.4. The second part of the induction step differs from the first part only by the length of the state contexts i that are considered for a transition from the current state to a next state. The termination step computes the desired likelihood of an emission sequence $\vec{o}(k)$ under an inhomogeneous $HHMM(L,C)$ λ with respect to the given transition class sequence $\vec{c}(k)$. This is done by marginalizing over the product of initial state probability π_{i_1}, emission density $b_{i_1}(o_1(k))$, and Backward-Variable $\beta_1^k(i)$ for each state context $i = (i_1) \in S^1$. The Backward algorithm has the same total run-time of $O\left((T_k - L)N^{L+1}\right)$ like the Forward algorithm. The run-time follows from the first part of the induction step that considers $T_k - L$ time steps in which the computation of one of the N^L Backward-Variables for one time step requires $O(N)$ operations. Good introductions to the Backward algorithm for a homogeneous HMM are given by Rabiner (1989), Durbin et al. (1998), and Bishop (2006).

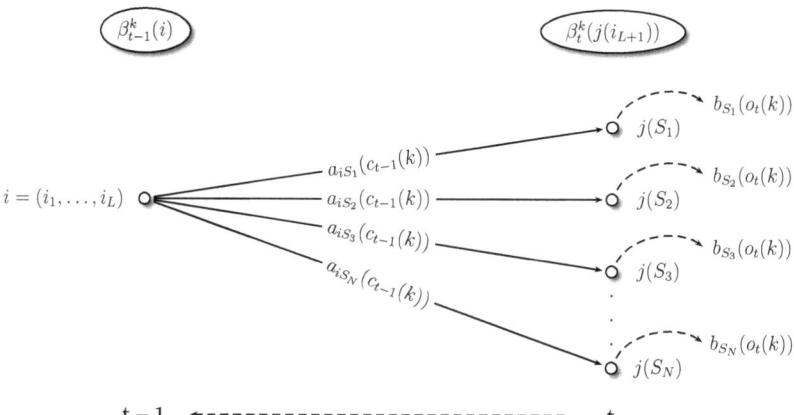

Figure 3.4: Computational scheme of the Backward-Variable $\beta_{t-1}^k(i)$ of state context $i = (i_1, \ldots, i_L) \in S^L$ during the first part of the induction step. Each Backward-Variable $\beta_t^k(j(i_{L+1}))$ of state context $j(i_{L+1})$ with $i_{L+1} \in S$ is considered by transforming i to $j(i_{L+1})$ via the transition from the current state i_L to the next state i_{L+1} using the corresponding transition probability $a_{ii_{L+1}}(c_{t-1}(k))$. After this transition the emission $o_t(k)$ is made by the state i_{L+1} with respect to its emission density $b_{i_{L+1}}(o_t(k))$.

3. Hidden Markov Models

3.3.3 Forward-Backward Procedure

The Forward-Backward procedure is an efficient way to compute the likelihood of an emission sequence $\vec{o}(k)$ under an inhomogeneous $HHMM(L,C)$ based on pre-computed Forward-Variables (3.3) and Backward-Variables (3.4). The following derivation computes the likelihood with respect to a fixed time step $L \leq t \leq T_k$.

$$P[\vec{o}(k) \,|\, \vec{c}(k), \lambda]$$
$$= \sum_{i \in S^L} P[\vec{o}(k), \vec{Q}_{t-L+1...t} = i \,|\, \vec{c}(k), \lambda]$$
$$= \sum_{i \in S^L} P[\vec{o}_{1...t}(k), \vec{Q}_{t-L+1...t} = i \,|\, \vec{c}_{1...t-1}(k), \lambda] \cdot P[\vec{o}_{t+1...T_k}(k) \,|\, \vec{Q}_{t-L+1...t} = i, \vec{c}_{t...T_k-1}(k), \lambda]$$
$$= \sum_{i \in S^L} \alpha_t^k(i) \cdot \beta_t^k(i) \tag{3.5}$$

The generalization to time steps $1 \leq t < L$ is straightforward by summing over $i \in S^t$ instead of $i \in S^L$ and by changing $\vec{Q}_{t-L+1...t}$ to $\vec{Q}_{1...t}$.

3.4 Solving the Optimal State Sequence Problem

The computation of an optimal state sequence $\vec{q}(k) = (q_1(k), \ldots, q_{T_k}(k))$ for an emission sequence $\vec{o}(k) = (o_1(k), \ldots, o_{T_k}(k))$ under an inhomogeneous $HHMM(L,C)$ λ with respect to a given transition class sequence $\vec{c}(k) = (c_1(k), \ldots, c_{T_k-1}(k))$ requires the definition of an optimality criterion. For that reason, the two most frequently used criteria, the State Posterior Criterion and the Viterbi Criterion, for determining an optimal state sequence under a homogeneous HMM are generalized for an inhomogeneous $HHMM(L,C)$.

3.4.1 State-Posterior Algorithm

The optimality criterion of the State-Posterior algorithm is to choose the most probable state $q_t(k) \in S$ for each time step $1 \leq t \leq T_k$ with respect to the given emission sequence $\vec{o}(k)$, the transition class sequence $\vec{c}(k)$, and the $HHMM(L,C)$ λ. The State-Posterior-Variable, also referred to as state-posterior, of state $i \in S$ at a fixed time step

$L \leq t \leq T_k$ is defined by

$$\gamma_t^k(i) := P[Q_t = i \mid \vec{o}(k), \vec{c}(k), \lambda]$$

$$= \frac{\sum_{j_1 \in S} \cdots \sum_{j_{L-1} \in S} \alpha_t^k((j_1, \ldots, j_{L-1}, i)) \cdot \beta_t^k((j_1, \ldots, j_{L-1}, i))}{\sum_{j \in S^L} \alpha_t^k(j) \cdot \beta_t^k(j)} \quad (3.6)$$

representing the probability of being in state $i \in S$ at time step t given the emission sequence $\vec{o}(k)$, the corresponding transition class sequence $\vec{c}(k)$, and the *HHMM(L,C)* λ. The State-Posterior-Variable $\gamma_t^k(i)$ is computed in terms of the Forward-Variable (3.3) and the Backward Variable (3.4). The Forward-Variable accounts for emitting the partial emission sequence $\vec{o}_{1...t}(k)$ and reaching state i in time step t, while the Backward-Variable accounts for the remaining partial emission sequence $\vec{o}_{t+1...T_k}(k)$ that is emitted after leaving state i in time step t. The focus is turned on state i by marginalizing over its memory $\vec{q}_{t-L+1...t-1} = (j_1, \ldots, j_{L-1})$ of predecessor states. The denominator given in (3.5) is the normalization factor to ensure that the state-posterior (3.6) represents a probability. The generalization to time steps $1 \leq t < L$ is straightforward by marginalizing over the memory of size $t-1$ instead of size $L-1$ for computing the numerator, and by changing the denominator to a sum over $j \in S^t$ instead of $j \in S^L$. Based on that, the most probable state

$$q_t(k) := \underset{i \in S}{\text{argmax}}\, \gamma_t^k(i)$$

is chosen for each time step $1 \leq t \leq T_k$. The run-time of the State-Posterior algorithm is $O\left((T_k - L)N^{L+1}\right)$. This follows from the computation of the Forward-Variables (3.3) and the Backward-Variables (3.4). The run-time reduces to $O\left((T_k - L + 1)N^L\right)$ if the Forward-Variables and Backward-Variables have already been computed. The State-Posterior algorithm is given by Rabiner (1989) and by Durbin et al. (1998) for a homogeneous *HMM*.

3.4.2 Viterbi Algorithm

The Viterbi algorithm is an efficient dynamic programming approach for computing the most probable state sequence $\vec{q}(k)$ that best explains the corresponding emission sequence $\vec{o}(k)$ given the transition class sequence $\vec{c}(k)$ and the inhomogeneous

3. Hidden Markov Models

$HHMM(L, C)$ λ. The computation of the so-called Viterbi path $\vec{q}(k)$ is done in three steps that require the definition of two variables. The first variable is the Viterbi-Variable

$$\delta_t^k(i) := \begin{cases} P[\vec{o}_{1...t}(k), \vec{Q}_{1...t} = i \,|\, \vec{c}_{1...t-1}(k), \lambda] & , 1 \leq t \leq L \\ \max_{w \in S^{t-L}} P[\vec{o}_{1...t}(k), \vec{Q}_{1...t-L} = w, \vec{Q}_{t-L+1...t} = i \,|\, \vec{c}_{1...t-1}(k), \lambda] & , L < t \leq T_k \end{cases}$$

(3.7)

which represents, in consideration of time step t, the joint probability density of the partial emission sequence $\vec{o}_{1...t}(k)$ and their corresponding most probable partial state sequence $\vec{q}_{1...t}$ ending with the state context $i \in S^{\min(t,L)}$. The second variable for time steps $L < t \leq T_k$ and each state context $i = (i_1, \ldots, i_L) \in S^L$ with $j(i_0) = (i_0, i_1, \ldots, i_{L-1})$ is the Backtrack-Variable

$$\Psi_t^k(i) := \underset{i_0 \in S}{\text{argmax}}\, \delta_{t-1}^k(j(i_0)) \cdot a_{j(i_0)i_L}(c_{t-1}(k)) \cdot b_{i_L}(o_t(k)) \qquad (3.8)$$

which stores the most probable L-th predecessor state $q_{t-L} = i_0 \in S$ of the current state $q_t = i_L \in S$. That is, the Backtrack-Variable contains the predecessor state that best explains the transition from state $q_{t-1} = i_{L-1}$ to state $q_t = i_L$ with respect to the memory $(i_0, i_1, \ldots, i_{L-2})$ of state i_{L-1}. Based on these two variables, the Viterbi path $\vec{q}(k)$ is computed by the following algorithm for an emission sequence $\vec{o}(k)$ of length $T_k \geq L$.

- Initialization

$$\forall i = (i_1) \in S^1: \quad \delta_1^k(i) = \pi_{i_1} \cdot b_{i_1}(o_1(k))$$

- Induction

$$\forall 1 \leq t < L \text{ and } \forall i = (i_1, \ldots, i_{t+1}) \in S^{t+1} \text{ with } j = (i_1, \ldots, i_t)$$

$$\delta_{t+1}^k(i) = \delta_t^k(j) \cdot a_{ji_{t+1}}(c_t(k)) \cdot b_{i_{t+1}}(o_{t+1}(k))$$

$$\forall L \leq t < T_k \text{ and } \forall i = (i_1, \ldots, i_L) \in S^L \text{ with } j(i_0) = (i_0, i_1, \ldots, i_{L-1})$$

$$\delta_{t+1}^k(i) = \max_{i_0 \in S} \delta_t^k(j(i_0)) \cdot a_{j(i_0)i_L}(c_t(k)) \cdot b_{i_L}(o_{t+1}(k))$$

$$\Psi_{t+1}^k(i) = \underset{i_0 \in S}{\text{argmax}}\, \delta_t^k(j(i_0)) \cdot a_{j(i_0)i_L}(c_t(k)) \cdot b_{i_L}(o_{t+1}(k))$$

- *Termination*
    1. Last L states

    $$\vec{q}_{T_k-L+1...T_k}(k) = \underset{i \in S^L}{\text{argmax}}\, \delta^k_{T_k}(i)$$

    2. Predecessor state $q_{T_k-t}(k)$ for all $L \leq t < T_k$

    $$q_{T_k-t}(k) = \Psi^k_{T_k-t+L}(\vec{q}_{T_k-t+1...T_k-t+L}(k))$$

The initialization step of the Viterbi algorithm is identical with the initialization step of the Forward algorithm. The induction step of the Viterbi algorithm has two main differences in comparison to the induction step of the Forward algorithm. First, the sum over all states $i_0 \in S$ is replaced by a maximization over all states $i_0 \in S$ to turn the focus of the Viterbi algorithm on one optimal state sequence. Second, the Backtrack-Variable is introduced to enable the computation of the Viterbi path $\vec{q}(k)$ by a backtracking procedure in the termination step. The computational scheme of the second part of the induction step is shown in Fig. 3.5. The run-time of the Viterbi algorithm is $O\left((T_k - L)N^{L+1}\right)$. This follows from the second part of the induction step that considers $T_k - L$ time steps in which the computation of one of the N^L Viterbi-Variables and of one of the N^L Backtrack-Variables requires $O(N)$ operations per time step. The Viterbi algorithm was initially introduced by Viterbi (1967). More recent descriptions are given by Rabiner (1989), Durbin et al. (1998), and Bishop (2006) for a homogeneous *HMM*. The extension of the Viterbi algorithm to a homogeneous *HHMM*(2) is given by He (1988), and the extension to a homogeneous *HHMM*(L) is described by Ching et al. (2003).

3. Hidden Markov Models

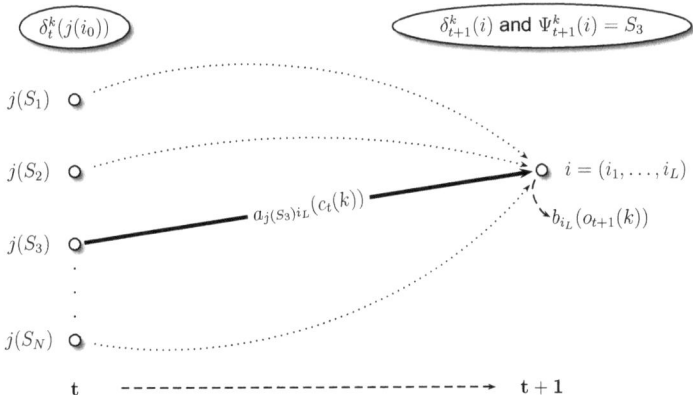

Figure 3.5: Computational scheme of the Viterbi-Variable $\delta_{t+1}^k(i)$ and the Backtrack-Variable $\Psi_{t+1}^k(i)$ for state context $i = (i_1, \ldots, i_L) \in S^L$ during the second part of the induction step. Each Viterbi-Variable $\delta_t^k(j(i_0))$ of state context $j(i_0) = (i_0, \ldots, i_{L-1})$ with $i_0 \in S$ is considered by transforming $j(i_0)$ to i based on the transition from the current state i_{L-1} to the next state i_L using the corresponding transition probability $a_{j(i_0)i_L}(c_t(k))$. After this transition the emission $o_{t+1}(k)$ is done by the state i_L with respect to its emission density $b_{i_L}(o_{t+1}(k))$. In contrast to the scheme shown for the Forward algorithm in Fig. 3.3, only the optimal link between one of all $\delta_t^k(j(i_0))$ and $\delta_{t+1}^k(i)$ is further considered. For instance, here the link between $\delta_t^k(j(S_3))$ and $\delta_{t+1}^k(i)$ is assumed to be optimal. This leads to the storage of $\Psi_{t+1}^k(i) = S_3$. This value is required for in the backtracking step of the Viterbi algorithm if the Viterbi path goes through the state context i at time step $t+1$.

3.5 Solving the Maximum Likelihood Problem

The process of adjusting the initial state probabilities, the transition probabilities, and the emission parameters of an inhomogeneous $HHMM(L, C)$ is called training. The objective function for solving the *Maximum Likelihood Problem* is the likelihood

$$P[\vec{o}(1), \ldots, \vec{o}(K) \mid \vec{c}(1), \ldots, \vec{c}(K), \lambda] = \prod_{k=1}^{K} P[\vec{o}(k) \mid \vec{c}(k), \lambda] \qquad (3.9)$$

of K statistically independent emission sequences $\vec{o}(1), \ldots, \vec{o}(K)$ under an inhomogeneous $HHMM(L, C)$ λ with respect to the given transition class sequences $\vec{c}(1), \ldots, \vec{c}(K)$. Due to the fact that the hidden state sequences must be considered in the computation of the likelihood (3.9), no analytical way is known that directly determines initial state probabilities, transition probabilities, and emission parameters of the $HHMM(L, C)$ λ for maximizing the likelihood. In this situation, the Baum-Welch algorithm published in a series of articles (Baum and Eagen (1967); Baum et al. (1970); Baum (1972)) is applied for maximizing the log-likelihood of (3.9) iteratively. The Baum-Welch algorithm is a special case of *Expectation Maximization* algorithms (EM algorithms) described by Dempster et al. (1977) that deal with hidden data to provide an effective local optimization procedure. More recent introductions to the Baum-Welch algorithm for a homogeneous HMM are given by Rabiner (1989), Durbin et al. (1998), and Bishop (2006). The basis of the Baum-Welch algorithm is given by Baum's auxiliary function

$$Q(\lambda \mid \lambda(h)) := \sum_{k=1}^{K} \sum_{\vec{q} \in S^{T_k}} P[\vec{q} \mid \vec{o}(k), \vec{c}(k), \lambda(h)] \cdot \log\left(P[\vec{o}(k), \vec{q} \mid \vec{c}(k), \lambda]\right) \qquad (3.10)$$

which is used to locally maximize the log-likelihood of (3.9) in consideration of the current $HHMM(L, C)$ $\lambda(h)$ of iteration step h. The rest of this section focuses on the extension of the Baum-Welch algorithm to the inhomogeneous $HHMM(L, C)$.

3.5.1 Baum-Welch Algorithm

The Baum-Welch algorithm is an iterative training procedure to locally maximize the log-likelihood of (3.9) by dealing with each hidden state sequence $\vec{q} = (q_1, \ldots, q_{T_k})$ that is able to emit a given emission sequence $\vec{o}(k) = (o_1(k), \ldots, o_{T_k}(k))$. The contribution of a state sequence $\vec{q}$ to the likelihood of the emission sequence $\vec{o}(k)$ is expressed by

3. Hidden Markov Models

the complete-data likelihood

$$P[\vec{o}(k), \vec{q}\,|\,\vec{c}(k), \lambda] = \pi_{q_1} \prod_{t=1}^{L-1} a_{\vec{q}_{1\ldots t} q_{t+1}}(c_t(k)) \prod_{t=L}^{T_k-1} a_{\vec{q}_{t-L+1\ldots t} q_{t+1}}(c_t(k)) \prod_{t=1}^{T_k} b_{q_t}(o_t(k)) \quad (3.11)$$

under the inhomogeneous *HHMM*(L, C) λ with respect to the given transition class sequence $\vec{c}(k)$. The relation $P[\vec{o}(k), \vec{q}\,|\,\vec{c}(k), \lambda] = P[\vec{q}\,|\,\vec{o}(k), \vec{c}(k), \lambda] \cdot P[\vec{o}(k)\,|\,\vec{c}(k), \lambda]$ between the complete-data likelihood and the likelihood of emission sequence $\vec{o}(k)$ is used to derive an alternative expression of the log-likelihood

$$\log(P[\vec{o}(k)\,|\,\vec{c}(k), \lambda]) = \log(P[\vec{o}(k), \vec{q}\,|\,\vec{c}(k), \lambda]) - \log(P[\vec{q}\,|\,\vec{o}(k), \vec{c}(k), \lambda]) \quad (3.12)$$

of emission sequence $\vec{o}(k)$ under the *HHMM*(L, C) λ with respect to transition class sequence $\vec{c}(k)$. In addition to this, the current *HHMM*(L, C) $\lambda(h)$ of iteration step h with known initial state probabilities, transition probabilities, and emission parameters is used to deduce the iterative training procedure. Based on this, the log-likelihood (3.12) is first multiplied with $P[\vec{q}\,|\,\vec{o}(k), \vec{c}(k), \lambda(h)]$ and then marginalized over all possible state sequences $\vec{q} \in S^{T_k}$. This leads to an alternative expression of the log-likelihood

$$\begin{aligned}
\log(P[\vec{o}(k)\,|\,\vec{c}(k), \lambda]) \\
&= \sum_{\vec{q} \in S^{T_k}} P[\vec{q}\,|\,\vec{o}(k), \vec{c}(k), \lambda(h)] \cdot \log(P[\vec{o}(k), \vec{q}\,|\,\vec{c}(k), \lambda]) \\
&- \sum_{\vec{q} \in S^{T_k}} P[\vec{q}\,|\,\vec{o}(k), \vec{c}(k), \lambda(h)] \cdot \log(P[\vec{q}\,|\,\vec{o}(k), \vec{c}(k), \lambda]) \quad (3.13)
\end{aligned}$$

under the next inhomogeneous *HHMM*(L, C) λ in terms of the current *HHMM*(L, C) $\lambda(h)$. The alternative expression of the log-likelihood given in (3.13) is used to rewrite the logarithmic right-hand side of the likelihood given in (3.9). This leads to an alternative expression of the objective function for solving the *Maximum Likelihood Problem*. This alternative expression of the objective function is the log-likelihood

3. Hidden Markov Models

$$\sum_{k=1}^{K} \log(P[\vec{o}(k) | \vec{c}(k), \lambda])$$

$$= \sum_{k=1}^{K} \sum_{\vec{q} \in S^{T_k}} P[\vec{q} | \vec{o}(k), \vec{c}(k), \lambda(h)] \cdot \log\left(P[\vec{o}(k), \vec{q} | \vec{c}(k), \lambda]\right)$$

$$- \sum_{k=1}^{K} \sum_{\vec{q} \in S^{T_k}} P[\vec{q} | \vec{o}(k), \vec{c}(k), \lambda(h)] \cdot \log\left(P[\vec{q} | \vec{o}(k), \vec{c}(k), \lambda]\right)$$

$$= Q(\lambda | \lambda(h)) - \sum_{k=1}^{K} \sum_{\vec{q} \in S^{T_k}} P[\vec{q} | \vec{o}(k), \vec{c}(k), \lambda(h)] \cdot \log\left(P[\vec{q} | \vec{o}(k), \vec{c}(k), \lambda]\right) \quad (3.14)$$

under the next inhomogeneous $HHMM(L, C)$ λ based on Baum's auxiliary function defined in (3.10) and the current inhomogeneous $HHMM(L, C)$ $\lambda(h)$ of iteration step h. To locally improve the log-likelihood under the next inhomogeneous $HHMM(L, C)$ λ in comparison to the log-likelihood under the current inhomogeneous $HHMM(L, C)$ $\lambda(h)$, the difference of both log-likelihoods

$$\sum_{k=1}^{K} \log(P[\vec{o}(k) | \vec{c}(k), \lambda]) - \sum_{k=1}^{K} \log(P[\vec{o}(k) | \vec{c}(k), \lambda(h)])$$

$$= Q(\lambda | \lambda(h)) - Q(\lambda(h) | \lambda(h))$$

$$+ \sum_{k=1}^{K} \sum_{\vec{q} \in S^{T_k}} P[\vec{q} | \vec{o}(k), \vec{c}(k), \lambda(h)] \cdot \log\left(\frac{P[\vec{q} | \vec{o}(k), \vec{c}(k), \lambda(h)]}{P[\vec{q} | \vec{o}(k), \vec{c}(k), \lambda]}\right) \quad (3.15)$$

must be positive. The last term on the right-hand side of (3.15) is a sum over relative entropies. The relative entropy is known to be always non-negative (Durbin et al. (1998); Bishop (2006)). For that reason, this sum only accounts for an improvement of the log-likelihood under the next inhomogeneous $HHMM(L, C)$ λ with respect to the current $HHMM(L, C)$ $\lambda(h)$. Thus, this sum can be neglected to simplify the computation of the parameters of the next inhomogeneous $HHMM(L, C)$ λ. In addition to this, Baum's auxiliary function $Q(\lambda(h) | \lambda(h))$ of the current inhomogeneous $HHMM(L, C)$ $\lambda(h)$ is independent of the new initial state probabilities, transition probabilities, and emission parameters of the next inhomogeneous $HHMM(L, C)$ λ. That is, for the computation of the parameters of the next inhomogeneous $HHMM(L, C)$ λ only Baum's auxiliary function $Q(\lambda | \lambda(h))$ on the right-hand side of (3.15) needs to be considered.

3. Hidden Markov Models

Based on this, the new parameters of the next inhomogeneous $HHMM(L, C)$

$$\lambda(h+1) = \underset{\lambda}{argmax}\, Q(\lambda \,|\, \lambda(h))$$

are computed by maximizing Baum's auxiliary function (3.10) over all possible initial state probabilities, transition probabilities, and emission parameters. Thus, iteratively choosing the parameters of the next inhomogeneous $HHMM(L, C)$ $\lambda(h+1)$ with respect to the current inhomogeneous $HHMM(L, C)$ $\lambda(h)$ leads to a positive difference of the log-likelihoods in (3.15) until a local maximum is reached. If a maximum is reached, the parameters of the next inhomogeneous $HHMM(L, C)$ $\lambda(h+1)$ are identical to those of the current inhomogeneous $HHMM(L, C)$ $\lambda(h)$. Thus, the log-likelihood does not change anymore. The convergence of the Baum-Welch algorithm to a local optimum of the likelihood has been proven in a series of articles (Baum and Eagen (1967); Baum et al. (1970); Baum (1972)). The proof of the local convergence for the general concept of EM algorithms, which includes the Baum-Welch algorithm as a special case, has been given by Dempster et al. (1977). Subsequently, the focus is on separating Baum's auxiliary function $Q(\lambda \,|\, \lambda(h))$ into three functions that enable the independent estimation of initial state probabilities, transition probabilities, and emission parameters of the next inhomogeneous $HHMM(L, C)$ $\lambda(h+1)$ with respect to the current inhomogeneous $HHMM(L, C)$ $\lambda(h)$.

3.5.2 Separating Baum's Auxiliary Function Into Parameter Classes

The complete-data likelihood $P[\vec{o}(k), \vec{q}\,|\,\vec{c}(k), \lambda]$ contained in Baum's auxiliary function $Q(\lambda \,|\, \lambda(h))$ in (3.10) can be substituted by their expression given in (3.11) to obtain independent functions for the parameters of the inhomogeneous $HHMM(L, C)$ λ. This leads to an alternative expression for $Q(\lambda \,|\, \lambda(h))$ given by

$$Q(\lambda \,|\, \lambda(h)) = Q_1(\vec{\pi} \,|\, \lambda(h)) + \left(\sum_{t=1}^{L} Q_2^t(A \,|\, \lambda(h)) \right) + Q_3(B \,|\, \lambda(h)) \qquad (3.16)$$

based on the separation into three independent functions for the parameters of the next $HHMM(L, C)$ λ. The function $Q_1(\vec{\pi} \,|\, \lambda(h))$ represents the initial state probabilities, $Q_2^t(A \,|\, \lambda(h))$ the transition probabilities, and $Q_3(B \,|\, \lambda(h))$ represents the emission parameters. In the following, these three functions are investigated in detail to provide the

3. Hidden Markov Models

basics to solve the *Maximum Likelihood Problem* using the Baum-Welch algorithm.

Initial State Parameters

Baum's auxiliary function for estimating the initial state probabilities π_i of the next inhomogeneous $HHMM(L,C)$ $\lambda(h+1)$ is given by

$$Q_1(\vec{\pi} \mid \lambda(h)) := \sum_{k=1}^{K} \sum_{\vec{q} \in S^{T_k}} P[\vec{q} \mid \vec{o}(k), \vec{c}(k), \lambda(h)] \cdot \log(\pi_{q_1})$$

$$= \sum_{i \in S} \log(\pi_i) \sum_{k=1}^{K} P[Q_1 = i \mid \vec{o}(k), \vec{c}(k), \lambda(h)]$$

$$= \sum_{i \in S} \Lambda_{\pi_i} \sum_{k=1}^{K} \gamma_1^k(i) \qquad (3.17)$$

based on expressing the sum over all state sequences $\vec{q} \in S^{T_k}$ by two sums. The first sum considers all initial states $i \in S$, and the second sum marginalizes over all state sequences $\vec{q} \in S^{T_k}$ with initial state $q_1 = i$ leading to $P[Q_1 = i \mid \vec{o}(k), \vec{c}(k), \lambda(h)]$, which represents the state posterior $\gamma_1^k(i)$ defined in (3.6) under the current inhomogeneous $HHMM(L,C)$ $\lambda(h)$. Finally, the logarithmic initial state probability $\log(\pi_i)$ is parameterized by $\Lambda_{\pi_i} := \log(\pi_i)$ to provide the basics for the parameter estimation.

Transition Parameters

For the estimation of the transition parameters $a_{ij}(c)$ of the next inhomogeneous $HHMM(L,C)$ $\lambda(h+1)$ one has to account for the expected number of transitions under the current inhomogeneous $HHMM(L,C)$ $\lambda(h)$. On that score, the Epsilon-Variable

$$\varepsilon_t^k(i,j) := \begin{cases} P[\vec{Q}_{1\ldots t} = i, Q_{t+1} = j \mid \vec{o}(k), \vec{c}(k), \lambda(h)] & , 1 \leq t < L \\ P[\vec{Q}_{t-L+1\ldots t} = i, Q_{t+1} = j \mid \vec{o}(k), \vec{c}(k), \lambda(h)] & , L \leq t < T_k \end{cases} \qquad (3.18)$$

defines the probability of having the state context $i \in S^{\min(t,L)}$ with $i = (i_1, \ldots, i_{\min(t,L)})$ at time step t and being in the next state $j \in S$ at time step $t+1$ given the emission sequence $\vec{o}(k)$, the transition class sequence $\vec{c}(k)$, and the current inhomogeneous $HHMM(L,C)$ $\lambda(h)$. The Epsilon-Variable defined in (3.18) generalizes the definition of the Epsilon-Variable given for a homogeneous *HMM* by Rabiner (1989). The computation of the Epsilon-Variable is done by utilizing the Forward-Variable (3.3) and the

3. Hidden Markov Models

Backward-Variable (3.4). The different Epsilon-Variables of a time step $1 \leq t < L$ are computed by

$$\varepsilon_t^k(i,j) = \frac{\alpha_t^k(i) \cdot a_{ij}(c_t(k)) \cdot b_j(o_{t+1}(k)) \cdot \beta_{t+1}^k((i_1, \ldots, i_t, j))}{\sum_{v:=(v_1,\ldots,v_t) \in S^t} \sum_{w \in S} \alpha_t^k(v) \cdot a_{vw}(c_t(k)) \cdot b_w(o_{t+1}(k)) \cdot \beta_{t+1}^k((v_1, \ldots, v_t, w))} \quad (3.19)$$

taking into account the transition from the current state i_t to next state j in consideration of the memory $(i_1, \ldots, i_{t-1})$ of the current state, and by integrating the emission $o_{t+1}(k)$ done by the next state j. In addition to this, all different Epsilon-Variables of a time step $L \leq t < T_k$ are computed similarly by

$$\varepsilon_t^k(i,j) = \frac{\alpha_t^k(i) \cdot a_{ij}(c_t(k)) \cdot b_j(o_{t+1}(k)) \cdot \beta_{t+1}^k((i_2, \ldots, i_L, j))}{\sum_{v:=(v_1,\ldots,v_L) \in S^L} \sum_{w \in S} \alpha_t^k(v) \cdot a_{vw}(c_t(k)) \cdot b_w(o_{t+1}(k)) \cdot \beta_{t+1}^k((v_2, \ldots, v_L, w))} \quad (3.20)$$

using the full memory of size L. That is, the Backward-Variable of time step $t+1$ does not account for the L-th predecessor state $q_{t-L+1} = i_1$ or $q_{t-L+1} = v_1$.

Based on the Epsilon-Variables (3.19) defined for the current inhomogeneous $HHMM(L,C)$ $\lambda(h)$, Baum's auxiliary function for estimating the transition probabilities of the next inhomogeneous $HHMM(L,C)$ $\lambda(h+1)$ is given by

$$Q_2^t(A \mid \lambda(h)) := \sum_{k=1}^K \sum_{\vec{q} \in S^{T_k}} P[\vec{q} \mid \vec{o}(k), \vec{c}(k), \lambda(h)] \cdot \log\left(a_{\vec{q}_{1\ldots t}q_{t+1}}(c_t(k))\right)$$

$$= \sum_{c \in \mathcal{C}} \sum_{i \in S^t} \sum_{j \in S} \log(a_{ij}(c)) \sum_{\substack{k=1 \\ c_t(k)=c}}^K P[\vec{Q}_{1\ldots t} = i, Q_{t+1} = j \mid \vec{o}(k), \vec{c}(k), \lambda(h)]$$

$$= \sum_{c \in \mathcal{C}} \sum_{i \in S^t} \sum_{j \in S} \Lambda_{a_{ij}(c)} \sum_{\substack{k=1 \\ c_t(k)=c}}^K \varepsilon_t^k(i,j) \quad (3.21)$$

for time steps $1 \leq t < L$. Here, the sum over all state sequences $\vec{q} \in S^{T_k}$ is substituted by four sums. Three of these sums are shown explicitly and the fourth sum can be substituted as explained subsequently. The first sum considers all transition classes $c \in \mathcal{C}$. The second sum considers all state contexts $i \in S^t$. The third sum considers all next states $j \in S$. Now, a fourth sum is necessary to marginalize over all state sequences $\vec{q} \in S^{T_k}$ with fixed state context $\vec{q}_{1\ldots t} = i$ and fixed next state $q_{t+1} = j$ in consideration that the transition class c of the first sum is identical to the given transition class $c_t(k)$. This fourth sum can be simplified to its marginal distribution $P[\vec{Q}_{1...t} = i, Q_{t+1} = j \,|\, \vec{o}(k), \vec{c}(k), \lambda(h)\,]$, which represents the Epsilon-Variable $\varepsilon_t^k(i,j)$ defined in (3.18). Finally, to provide the basics for the parameter estimation, the logarithmic transition probability $\log(a_{ij}(c))$ is parameterized by $\Lambda_{a_{ij}(c)} := \log(a_{ij}(c))$. In analogy to this, Baum's auxiliary function to estimate the transition probabilities used at time steps $t \geq L$ is defined by

$$Q_2^L(A \,|\, \lambda(h)) := \sum_{k=1}^{K} \sum_{t=L}^{T_k-1} \sum_{\vec{q} \in S^{T_k}} P[\vec{q} \,|\, \vec{o}(k), \vec{c}(k), \lambda(h)\,] \cdot \log\left(a_{\vec{q}_{t-L+1}...t q_{t+1}}(c_t(k))\right)$$

$$= \sum_{c \in \mathcal{C}} \sum_{i \in S^L} \sum_{j \in S} \Lambda_{a_{ij}(c)} \sum_{k=1}^{K} \sum_{\substack{t=L \\ \mathbf{c_t(k)} = \mathbf{c}}}^{T_k-1} \varepsilon_t^k(i,j) \qquad (3.22)$$

for the next inhomogeneous $HHMM(L, C)$ $\lambda(h+1)$ with respect to the Epsilon-Variable $\varepsilon_t^k(i,j)$ defined in (3.20). The derivation is nearly identical to that of $Q_2^t(A \,|\, \lambda(h))$ given in (3.21). The only difference results from an additional sum over all time steps $L \leq t < T_k$.

Emission Parameters

Baum's auxiliary function to estimate the emission parameters of the next inhomogeneous $HHMM(L, C)$ $\lambda(h+1)$ is given by

$$Q_3(B \,|\, \lambda(h)) := \sum_{k=1}^{K} \sum_{t=1}^{T_k} \sum_{\vec{q} \in S^{T_k}} P[\vec{q} \,|\, \vec{o}(k), \vec{c}(k), \lambda(h)\,] \cdot \log(b_{q_t}(o_t(k)))$$

$$= \sum_{i \in S} \sum_{k=1}^{K} \sum_{t=1}^{T_k} \log(b_i(o_t(k))) \cdot P[Q_t = i \,|\, \vec{o}(k), \vec{c}(k), \lambda(h)\,]$$

$$= \sum_{i \in S} \sum_{k=1}^{K} \sum_{t=1}^{T_k} \log(b_i(o_t(k))) \cdot \gamma_t^k(i) \qquad (3.23)$$

including the substitution of the sum over all state sequences $\vec{q} \in S^{T_k}$ by two sums. The first sum runs over all current states $i \in S$. Now, a second sum is required to marginalize over all state sequences $\vec{q} \in S^{T_k}$ with fixed current state $q_t = i$. The second sum can be simplified to its marginal probability $P[Q_t = i \,|\, \vec{o}(k), \vec{c}(k), \lambda(h)\,]$, which is exactly the state posterior $\gamma_t^k(i)$ defined in (3.6) computed for the current inhomogeneous $HHMM(L, C)$ $\lambda(h)$.

3. Hidden Markov Models

3.5.3 Estimating HHMM Parameters

The estimation of the initial state probabilities $\pi_i^{(h+1)}$ and the transition probabilities $a_{ij}(c)^{(h+1)}$ of the next inhomogeneous $HHMM(L,C)$ $\lambda(h+1)$ is done by using the standard Lagrange optimization technique described in the textbook by Bishop (2006). An outline to the proofs that the obtained re-estimation formulas for $\pi_i^{(h+1)}$ and $a_{ij}(c)^{(h+1)}$ maximize their corresponding Baum's auxiliary function is given by Durbin et al. (1998) for a homogeneous HMM. The generalization of these proofs to an inhomogeneous $HHMM(L,C)$ is straightforward. For that reason, only the parameter estimation formulas are provided.

Initial State Parameters

To determine the initial state distribution $\vec{\pi}(h+1)$ of the next inhomogeneous $HHMM(L,C)$ $\lambda(h+1)$ one has to maximize Baum's auxiliary function $Q_1(\vec{\pi} \mid \lambda(h))$ given in (3.17) in subject to the constraint $\sum_{i \in S} \exp(\Lambda_{\pi_i}) = 1$. This is done by using the auxiliary function $Q_1(\vec{\pi} \mid \lambda(h)) - \delta \cdot ((\sum_{i \in S} \exp(\Lambda_{\pi_i})) - 1)$ with Lagrange multiplier δ. The auxiliary function is differentiated with respect to Λ_{π_i} and δ. Both derivatives are set equal to zero to compute the initial state probability. Under consideration of the relation $\pi_i = \exp(\Lambda_{\pi_i})$, the initial state probability

$$\pi_i^{(h+1)} = \frac{\sum_{k=1}^{K} \gamma_1^k(i)}{\sum_{v \in S} \sum_{k=1}^{K} \gamma_1^k(v)} \quad (3.24)$$

for state $i \in S$ under the next inhomogeneous $HHMM(L,C)$ $\lambda(h+1)$ is obtained with respect to all state posteriors $\gamma_1^k(i)$ and $\gamma_1^k(v)$ given in (3.6) for the current inhomogeneous $HHMM(L,C)$ $\lambda(h)$.

Transition Parameters

The transition probability $a_{ij}(c)^{(h+1)}$ used at a fixed time step $1 \leq t < L$ by the next inhomogeneous $HHMM(L,C)$ $\lambda(h+1)$ is determined by maximizing Baum's auxiliary function $Q_2^t(A \mid \lambda(h))$ given in (3.21) in subject to the constraint $\sum_{j \in S} \exp(\Lambda_{a_{ij}(c)}) = 1$. This is done based on the auxiliary function $Q_2^t(A \mid \lambda(h)) - \sum_{i \in S^t} \delta_i \cdot ((\sum_{j \in S} \exp(\Lambda_{a_{ij}(c)})) - 1)$ with Lagrange multiplier δ_i. The auxiliary function is differentiated with respect to

3. Hidden Markov Models

$\Lambda_{a_{ij}(c)}$ and δ_i. Both derivatives are set equal to zero to compute the transition probability. Based on the relation $a_{ij}(c) = \exp(\Lambda_{a_{ij}(c)})$, one obtains for each state context $i = (i_1, \ldots, i_t) \in S^t$ and each next state $j \in S$ the transition probability

$$a_{ij}(c)^{(h+1)} = \frac{\sum_{\substack{k=1 \\ c_t(k) = c}}^{K} \varepsilon_t^k(i,j)}{\sum_{v \in S} \sum_{\substack{k=1 \\ c_t(k) = c}}^{K} \varepsilon_t^k(i,v)} \qquad (3.25)$$

for a transition from the current state i_t to the next state j in transition class $c \in \mathcal{C}$ with respect to the memory $(i_1, \ldots, i_{t-1})$ of the current state. The Epsilon-Variables $\varepsilon_t^k(i,j)$ and $\varepsilon_t^k(i,v)$ given in (3.19) that are necessary for the estimation of this transition probability have to be computed under the current inhomogeneous HHMM(L, C) $\lambda(h)$.
Similar to this, all transition probabilities for time steps $t \geq L$ are determined by maximizing Baum's auxiliary function $Q_2^L(A \mid \lambda(h))$ given in (3.22) in subject to the constraint $\sum_{j \in S} \exp(\Lambda_{a_{ij}(c)}) = 1$ using the auxiliary function $Q_2^L(A \mid \lambda(h)) - \sum_{i \in S^L} \delta_i \cdot ((\sum_{j \in S} \exp(\Lambda_{a_{ij}(c)})) - 1)$ with Lagrange multiplier δ_i. In analogy, one obtains for each state context $i = (i_1, \ldots, i_L) \in S^L$ and each next state $j \in S$ the transition probability

$$a_{ij}(c)^{(h+1)} = \frac{\sum_{k=1}^{K} \sum_{\substack{t=L \\ c_t(k) = c}}^{T_k-1} \varepsilon_t^k(i,j)}{\sum_{v \in S} \sum_{k=1}^{K} \sum_{\substack{t=L \\ c_t(k) = c}}^{T_k-1} \varepsilon_t^k(i,v)} \qquad (3.26)$$

for a transition from the current state i_L to the next state j in transition class $c \in \mathcal{C}$ in consideration of the memory $(i_1, \ldots, i_{L-1})$ of the current state. The Epsilon-Variables $\varepsilon_t^k(i,j)$ and $\varepsilon_t^k(i,v)$ defined in (3.20) that are required for the computation of this transition probability are computed under the current inhomogeneous HHMM(L,C) $\lambda(h)$.

Emission Parameters

Each state $i \in S$ of the inhomogeneous HHMM(L,C) is characterized by a Gaussian emission density $b_i(o)$ defined in (3.2) with a state-specific mean μ_i and a state-specific standard deviation σ_i. To estimate the mean $\mu_i^{(h+1)}$ and the standard deviation $\sigma_i^{(h+1)}$

3. Hidden Markov Models

for the next inhomogeneous $HHMM(L,C)$ $\lambda(h+1)$ one has to maximize $Q_3(B \mid \lambda(h))$ given in (3.23). This is done by determining the critical points for which the derivatives of $Q_3(B \mid \lambda(h))$ for the mean μ_i and the standard deviation σ_i are identical to zero. This leads to the state-specific mean

$$\mu_i^{(h+1)} = \frac{\sum_{k=1}^{K} \sum_{t=1}^{T_k} o_t(k) \cdot \gamma_t^k(i)}{\sum_{k=1}^{K} \sum_{t=1}^{T_k} \gamma_t^k(i)} \qquad (3.27)$$

and the state-specific standard deviation

$$\sigma_i^{(h+1)} = \sqrt{\frac{\sum_{k=1}^{K} \sum_{t=1}^{T_k} \left(o_t(k) - \mu_i^{(h+1)}\right)^2 \cdot \gamma_t^k(i)}{\sum_{k=1}^{K} \sum_{t=1}^{T_k} \gamma_t^k(i)}} \qquad (3.28)$$

of state $i \in S$ for the next inhomogeneous $HHMM(L,C)$ $\lambda(h+1)$. Both, the mean and the standard deviation require the state posteriors $\gamma_t^k(i)$ given in (3.6) that have to be computed under the current inhomogeneous $HHMM(L,C)$ $\lambda(h)$. For further details, the introduction to the estimation of the mean and the standard deviation by Bilmes (1998) can be considered.

3.5.4 Computational Scheme of the Baum-Welch Algorithm

The computational scheme of the Baum-Welch algorithm is specified subsequently in terms of an initialization and an iteration step under consideration of the basics obtained in the previous sections.

- *Initialization*: Choose initial state probabilities, transition probabilities, and emission parameters of the inhomogeneous $HHMM(L,C)$ $\lambda(1)$.

- *Iteration*: For iteration steps $h = 1, 2, \ldots$
  - Use the current inhomogeneous $HHMM(L,C)$ $\lambda(h)$ to compute all State-Posterior-Variables $\gamma_t^k(i)$ given in (3.6) and all Epsilon-Variables $\varepsilon_t^k(i,j)$ given

in (3.19) and (3.20) based on the given emission sequences $\vec{o}(1), \ldots, \vec{o}(K)$ and their corresponding transition class sequences $\vec{c}(1), \ldots, \vec{c}(K)$.
- Compute the optimal parameters of the next inhomogeneous $HHMM(L, C)$ $\lambda(h + 1)$ based on the previous computations.
    1. Compute the initial state probability $\pi_i^{(h+1)}$ given in (3.24) for each state $i \in S$.
    2. Compute the transition probability $a_{ij}(c)^{(h+1)}$ for each state context $i \in S^t$ of length $1 \leq t < L$ and each next state $j \in S$ using (3.25). In analogy, use (3.26) to compute the transition probability for each state context $i \in S^L$ of length L.
    3. Compute the mean $\mu_i^{(h+1)}$ and the standard deviation $\sigma_i^{(h+1)}$ for each state $i \in S$ as specified in (3.27) and (3.28).
- Stop if the log-likelihood under the next inhomogeneous $HHMM(L, C)$ $\lambda(h+1)$ has increased less than a pre-defined threshold in comparison to the log-likelihood under the current inhomogeneous $HHMM(L, C)$ $\lambda(h)$, otherwise start the next iteration step with $h := h + 1$.

3.6 Prior

In Bayesian statistics, the prior represents a statistical distribution over the parameters of a model that allows to assign an *a priori* probability to each individual model; see e.g. Durbin et al. (1998) or Bishop (2006). This provides the opportunity to integrate biological prior knowledge into the estimation of the initial state probabilities, the transition probabilities, and the emission parameters of an inhomogeneous $HHMM(L, C)$ λ. The choice of specific prior distributions should also provide the basics for the analytical estimation of these parameters. This can be realized by the choice of conjugate priors that greatly simplify the parameter estimation (Bishop (2006)). With the choice of a conjugate prior for a model parameter, the posterior of this parameter has the same functional form as the prior of this parameter. For the parameters of the inhomogeneous $HHMM(L, C)$, the conjugate prior of the initial state distribution and of the transition distribution of a fixed state $i \in S$ with a fixed memory on its predecessor states is a Dirichlet distribution (Durbin et al. (1998)). Regarding the state-specific emission parameters, the conjugate prior for the mean of a Gaussian distribution is a

3. Hidden Markov Models

Gaussian distribution (Bishop (2006)), and the conjugate prior for the standard deviation of a Gaussian distribution is an Inverted-Gamma distribution (Evans et al. (2000)). Based on this, the prior of the inhomogeneous $HHMM(L, C)$ λ is defined by

$$P[\lambda] := D_1(\vec{\pi}\,|\,\Theta_1) \cdot \left(\prod_{t=1}^{L} D_2^t(A\,|\,\Theta_2)\right) \cdot D_3(B\,|\,\Theta_3) \qquad (3.29)$$

as a product of independent priors $D_1(\vec{\pi}\,|\,\Theta_1)$, $D_2^t(A\,|\,\Theta_2)$, and $D_3(B\,|\,\Theta_3)$ for the initial state distribution, the transition probabilities, and the emission parameters. These priors are specified subsequently.

3.6.1 Initial State Parameter Prior

The prior for the initial state distribution $\vec{\pi}$ with initial state probability $\pi_i = \exp(\Lambda_{\pi_i})$ for each state $i \in S$ is given by the transformed Dirichlet distribution

$$D_1(\vec{\pi}\,|\,\Theta_1) := Z(\Theta_1) \prod_{i \in S} \exp(\Lambda_{\pi_i} \cdot \vartheta_i) \qquad (3.30)$$

based on the vector of parameters $\Theta_1 := (\vartheta_{S_1}, \ldots, \vartheta_{S_N})$ with $\vartheta_i \in \mathbb{R}^+$ for each state $i \in S$, and the normalization constant $Z(\Theta_1) := \Gamma(\sum_{i \in S} \vartheta_i)/\prod_{i \in S} \Gamma(\vartheta_i)$ with Gamma function $\Gamma(x) = \int_0^\infty u^{x-1} \cdot \exp(-u)\, du$ for all $x \in \mathbb{R}^+$. This transformed Dirichlet distribution is specified in a general form by MacKay (1998).

3.6.2 Transition Parameter Prior

The prior of the transition parameters of state context length t is the product of transformed Dirichlet distributions

$$D_2^t(A\,|\,\Theta_2) := \prod_{c \in C} \prod_{i \in S^t} Z(\Theta_2^i(c)) \prod_{j \in S} \exp(\Lambda_{a_{ij}(c)} \cdot \vartheta_{ij}(c)) \qquad (3.31)$$

with respect to the relation $a_{ij}(c) = \exp(\Lambda_{a_{ij}(c)})$ and in consideration of the matrices of parameters $\Theta_2 := (\Theta_2(1), \ldots, \Theta_2(C))$. For each transition class $c \in C$ a matrix $\Theta_2(c) := (\Theta_2^i(c))_{i \in S^t}$ is defined based on the vector $\Theta_2^i(c) := (\vartheta_{iS_1}(c), \ldots, \vartheta_{iS_N}(c))$ with $\vartheta_{ij}(c) \in \mathbb{R}^+$ to represent the prior knowledge for a transition from state context i to next state j in consideration of transition class c. The normalization constant of each Dirichlet distribution is defined by $Z(\Theta_2^i(c)) := \Gamma(\sum_{j \in S} \vartheta_{ij}(c))/\prod_{j \in S} \Gamma(\vartheta_{ij}(c))$.

3.6.3 Emission Parameter Prior

The prior of the state-specific emission parameters is the product of independent Gaussian-Inverted-Gamma distributions

$$D_3(B \mid \Theta_3) := \prod_{i \in S} N(\mu_i \mid \eta_i, \sigma_i/\sqrt{\epsilon_i}) \cdot I_G(\sigma_i \mid r_i, \alpha_i) \tag{3.32}$$

as defined by Evans et al. (2000). The parameters of the emission prior are defined by the matrix $\Theta_3 := ((\eta_i, \epsilon_i, r_i, \alpha_i))_{i \in S}$ that contains the mean $\eta_i \in \mathbb{R}$, the scale parameter $\epsilon_i \in \mathbb{R}^+$, the shape parameter $r_i \in \mathbb{R}^+$, and the scale parameter $\alpha_i \in \mathbb{R}^+$. The Gaussian-Inverted-Gamma distribution of state $i \in S$ consists of a Gaussian distribution

$$N(\mu_i \mid \eta_i, \sigma_i/\sqrt{\epsilon_i}) := \frac{\sqrt{\epsilon_i}}{\sqrt{2\pi}\sigma_i} \cdot \exp\left(-\frac{\epsilon_i}{2} \cdot \left(\frac{\mu_i - \eta_i}{\sigma_i}\right)^2\right)$$

as prior for the state-specific mean μ_i with mean η_i and standard deviation $\sigma_i/\sqrt{\epsilon_i}$, and an Inverted-Gamma distribution

$$I_G(\sigma_i \mid r_i, \alpha_i) := \frac{2\alpha_i^{r_i}}{\Gamma(r_i)\sigma_i^{2r_i+1}} \cdot \exp\left(-\frac{\alpha_i}{\sigma_i^2}\right)$$

as prior of the state-specific standard deviation σ_i with shape parameter r_i and scale parameter α_i. The Inverted-Gamma distribution as prior of the standard deviation σ_i can be derived from the transformation of a Gamma distribution over a variable x with respect to the relation $x = 1/\sigma_i^2$ using the general theory behind the changes of variables of probability densities briefly summarized in the textbook of Durbin et al. (1998). The Gamma distribution itself represents the conjugate prior for the precision (inverse variance) of a Gaussian density (Bishop (2006)).

3.7 Solving the Maximum A Posteriori Problem

The objective function for solving the *Maximum A Posteriori Problem* is the posterior $P[\lambda \mid \vec{o}(1), \ldots, \vec{o}(K), \vec{c}(1), \ldots, \vec{c}(K)]$ of the inhomogeneous *HHMM*(L, C) λ given K statistically independent emission sequences $\vec{o}(1), \ldots, \vec{o}(K)$ and their corresponding transition class sequences $\vec{c}(1), \ldots, \vec{c}(K)$. Using the Bayes' theorem and the stated independence assumption for the emission sequences, the posterior can be transformed

3. Hidden Markov Models

into

$$P[\lambda \,|\, \vec{o}(1), \ldots, \vec{o}(K), \vec{c}(1), \ldots, \vec{c}(K)] = \frac{P[\lambda] \cdot \prod_{k=1}^{K} P[\vec{o}(k) \,|\, \vec{c}(k), \lambda]}{\prod_{k=1}^{K} P[\vec{o}(k), \vec{c}(k)]} \quad (3.33)$$

a product of the prior $P[\lambda]$ and the likelihood $\prod_{k=1}^{K} P[\vec{o}(k) \,|\, \vec{c}(k), \lambda]$ divided by the normalization constant $\prod_{k=1}^{K} P[\vec{o}(k), \vec{c}(k)]$. The normalization constant is independent of the specific parameters of the inhomogeneous $HHMM(L,C)$ λ. Thus, neglecting this normalization constant, the posterior is proportional to the product of the prior and the likelihood. Again, like for the *Maximum Likelihood Problem*, no analytical solution is known that directly determines initial state probabilities, transition probabilities, and emission parameters of the inhomogeneous $HHMM(L,C)$ λ for maximizing (3.33). However, in this situation one can make use of Baum's auxiliary function $Q(\lambda \,|\, \lambda(h))$ given in (3.10) as an alternative expression of the log-likelihood. This strategy leads to a Bayesian version of the Baum-Welch algorithm for the iterative maximization of the posterior.

3.7.1 Bayesian Baum-Welch Algorithm

The Bayesian version of the Baum-Welch algorithm is an iterative procedure to locally maximize the posterior given in (3.33) by considering each hidden state sequence $\vec{q}$ that is able to emit a given emission sequence $\vec{o}(k)$ in consideration of the corresponding transition class sequence $\vec{c}(k)$. To derive this iterative training procedure the log-posterior of (3.33) has to be considered. The log-posterior of the next inhomogeneous $HHMM(L,C)$ λ

$$\log\left(P[\lambda \,|\, \vec{o}(1), \ldots, \vec{o}(K), \vec{c}(1), \ldots, \vec{c}(K)]\right) = \log\left(P[\lambda]\right) + \sum_{k=1}^{K} \log\left(P[\vec{o}(k) \,|\, \vec{c}(k), \lambda]\right) - Z$$

is expressed in terms of the log-prior, the log-likelihood, and the log-normalization constant $Z = \sum_{k=1}^{K} \log\left(P[\vec{o}(k), \vec{c}(k)]\right)$. By substituting the log-likelihood with their alternative expression given in (3.14), the log-posterior can be transformed to the alternative

3. Hidden Markov Models

expression

$$\log(P[\lambda \mid \vec{o}(1), \ldots, \vec{o}(K), \vec{c}(1), \ldots, \vec{c}(K)])$$
$$= \log(P[\lambda]) - Z + Q(\lambda \mid \lambda(h))$$
$$- \sum_{k=1}^{K} \sum_{\vec{q} \in S^{T_k}} P[\vec{q} \mid \vec{o}(k), \vec{c}(k), \lambda(h)] \cdot \log(P[\vec{q} \mid \vec{o}(k), \vec{c}(k), \lambda])$$

which is depending on the current inhomogeneous $HHMM(L,C)$ $\lambda(h)$ of iteration step h. To locally improve the log-posterior of the next inhomogeneous $HHMM(L,C)$ λ with respect to the log-posterior under the current inhomogeneous $HHMM(L,C)$ $\lambda(h)$, the difference of both log-posteriors

$$\log(P[\lambda \mid \vec{o}(1), \ldots, \vec{o}(K), \vec{c}(1), \ldots, \vec{c}(K)]) - \log(P[\lambda(h) \mid \vec{o}(1), \ldots, \vec{o}(K), \vec{c}(1), \ldots, \vec{c}(K)])$$
$$= \log(P[\lambda]) - \log(P[\lambda(h)]) + Q(\lambda \mid \lambda(h)) - Q(\lambda(h) \mid \lambda(h))$$
$$+ \sum_{k=1}^{K} \sum_{\vec{q} \in S^{T_k}} P[\vec{q} \mid \vec{o}(k), \vec{c}(k), \lambda(h)] \cdot \log\left(\frac{P[\vec{q} \mid \vec{o}(k), \vec{c}(k), \lambda(h)]}{P[\vec{q} \mid \vec{o}(k), \vec{c}(k), \lambda]}\right) \quad (3.34)$$

must be positive. Baum's auxiliary function $Q(\lambda(h) \mid \lambda(h))$ and the log-prior $\log(P[\lambda(h)])$ are only depending on the current inhomogeneous $HHMM(L,C)$ $\lambda(h)$ of iteration step h. Due to that, both terms are constants that do not influence the estimation of the parameters of the next inhomogeneous $HHMM(L,C)$ λ. The last term on the right-hand side is again, as in the derivation of the Baum-Welch algorithm, the sum over K relative entropies. The relative entropy is known to be always non-negative (Durbin et al. (1998); Bishop (2006)). Thus, the sum over relative entropies only accounts for a local improvement of the log-posterior under the next inhomogeneous $HHMM(L,C)$ λ with respect to the current inhomogeneous $HHMM(L,C)$ $\lambda(h)$. On that score, this sum is neglected to simplify the estimation of the parameters of the next inhomogeneous $HHMM(L,C)$ λ. The parameters of the next inhomogeneous $HHMM(L,C)$

$$\lambda(h+1) = \underset{\lambda}{\text{argmax}}\,(Q(\lambda \mid \lambda(h)) + \log(P[\lambda]))$$

are computed by maximizing the sum of Baum's auxiliary function (3.10) and the log-prior of (3.29) over all possible initial state probabilities, transition probabilities, and emission parameters. This maximization always leads to a positive difference of the log-posteriors (3.34) until a maximum is reached. If a maximum is reached, then the

3. Hidden Markov Models

parameters of the next inhomogeneous $HHMM(L,C)$ $\lambda(h+1)$ are identical to those of the current inhomogeneous $HHMM(L,C)$ $\lambda(h)$. Thus, the log-posterior does not change anymore. The proof of the local convergence of the Bayesian Baum-Welch algorithm is included in the general concept of EM algorithms introduced by Dempster et al. (1977). Subsequent to this section, the focus is on estimating the initial state probabilities, the transition probabilities, and the emission parameters of the next inhomogeneous $HHMM(L,C)$ $\lambda(h+1)$.

3.7.2 Estimating HHMM Parameters

The basis of the parameter estimation is the separation of Baum's auxiliary function $Q(\lambda \mid \lambda(h))$. As shown in (3.16), Baum's auxiliary function can be divided into independent functions for each class of parameters of the inhomogeneous $HHMM(L,C)$ λ. That is, the initial state probabilities are represented by $Q_1(\vec{\pi} \mid \lambda(h))$ defined in (3.17), two functions $Q_2^t(A \mid \lambda(h))$ are defined in (3.21) and (3.22) for the transition probabilities, and the emission parameters are represented by $Q_3(B \mid \lambda(h))$ defined in (3.23). Each of these functions is now combined with the corresponding prior distribution to determine the parameters of the next inhomogeneous $HHMM(L,C)$. In analogy to the parameter estimation for the Baum-Welch algorithm, the standard Lagrange optimization technique described in the textbook of Bishop (2006) is used to determine the initial state probabilities and the transition probabilities of the next inhomogeneous $HHMM(L,C)$.

Initial State Parameters

To determine the initial state distribution $\vec{\pi}(h+1)$ of the next inhomogeneous $HHMM(L,C)$ $\lambda(h+1)$ one has to maximize Baum's auxiliary function $Q_1(\vec{\pi} \mid \lambda(h))$ given in (3.17) in combination with the prior $D_1(\vec{\pi} \mid \Theta_1)$ given in (3.30) in subject to the constraint $\sum_{i \in S} \exp(\Lambda_{\pi_i}) = 1$. For that reason, the auxiliary function $Q_1(\vec{\pi} \mid \lambda(h)) + \log(D_1(\vec{\pi} \mid \Theta_1)) - \delta \cdot ((\sum_{i \in S} \exp(\Lambda_{\pi_i})) - 1)$ with Lagrange multiplier δ is defined. This auxiliary function is differentiated with respect to Λ_{π_i} and δ. Both derivatives are set equal to zero to compute the initial state probability. Using the relation $\pi_i = \exp(\Lambda_{\pi_i})$, the initial state probability

$$\pi_i^{(h+1)} = \frac{\left(\sum_{k=1}^{K} \gamma_1^k(i)\right) + \vartheta_i}{\left(\sum_{v \in S} \sum_{k=1}^{K} \gamma_1^k(v)\right) + \left(\sum_{v \in S} \vartheta_v\right)} \quad (3.35)$$

is obtained for state $i \in S$ of the next inhomogeneous *HHMM(L, C)* $\lambda(h+1)$. The required state posteriors $\gamma_1^k(i)$ and $\gamma_1^k(v)$ given in (3.6) have to be computed under the current inhomogeneous *HHMM(L, C)* $\lambda(h)$. The initial state probability $\pi_i^{(h+1)}$ in (3.35) obtained for the Bayesian Baum-Welch algorithm is extended to that given in (3.24) for the Baum-Welch algorithm by additionally including the prior parameter ϑ_i to express prior knowledge on the initial state probability.

Transition Parameters

The transition probability $a_{ij}(c)^{(h+1)}$ used by the next inhomogeneous *HHMM(L, C)* $\lambda(h+1)$ at a time step $1 \leq t < L$ is estimated by maximizing Baum's auxiliary function $Q_2^t(A \mid \lambda(h))$ given in (3.21) in combination with the prior $D_2^t(A \mid \Theta_2)$ defined in (3.31) in subject to the constraint $\sum_{j \in S} \exp(\Lambda_{a_{ij}(c)}) = 1$. The auxiliary function $Q_2^t(A \mid \lambda(h)) + \log(D_2^t(A \mid \Theta_2)) - \sum_{i \in S^t} \delta_i \cdot ((\sum_{j \in S} \exp(\Lambda_{a_{ij}(c)})) - 1)$ with Lagrange multiplier δ_i provides the basics for this maximization. This function is differentiated with respect to $\Lambda_{a_{ij}(c)}$ and δ_i. Both derivatives are set equal to zero to compute the transition probability. Based on the relation $a_{ij}(c) = \exp(\Lambda_{a_{ij}(c)})$, one obtains for each state context $i = (i_1, \ldots, i_t) \in S^t$ and each next state $j \in S$ the transition probability

$$a_{ij}(c)^{(h+1)} = \frac{\left(\sum_{\substack{k=1 \\ c_t(k)=c}}^{K} \varepsilon_t^k(i,j)\right) + \vartheta_{ij}(c)}{\left(\sum_{v \in S} \sum_{\substack{k=1 \\ c_t(k)=c}}^{K} \varepsilon_t^k(i,v)\right) + \left(\sum_{v \in S} \vartheta_{iv}(c)\right)} \quad (3.36)$$

for a transition from the current state i_t to the next state j in transition class $c \in \mathcal{C}$ in consideration of the memory $(i_1, \ldots, i_{t-1})$ of the current state. All required Epsilon-Variables $\varepsilon_t^k(i, j)$ and $\varepsilon_t^k(i, v)$ defined in (3.19) have to be computed under the current inhomogeneous *HHMM(L, C)* $\lambda(h)$. The transition probability $a_{ij}(c)^{(h+1)}$ in (3.36) ob-

3. Hidden Markov Models

tained for the Bayesian Baum-Welch algorithm is extended to that given in (3.25) for the Baum-Welch algorithm by additionally including the prior parameter $\vartheta_{ij}(c)$ to express prior knowledge for this transition probability.

Similar to this parameter estimation, all transition probabilities used by the inhomogeneous $HHMM(L,C)$ at time steps $t \geq L$ are determined by maximizing Baum's auxiliary function $Q_2^L(A \mid \lambda(h))$ given in (3.22) in combination with the prior $D_2^L(A \mid \Theta_2)$ defined in (3.31) in subject to the constraint $\sum_{j \in S} \exp(\Lambda_{a_{ij}(c)}) = 1$. This is done by using the auxiliary function $Q_2^L(A \mid \lambda(h)) + \log(D_2^L(A \mid \Theta_2)) - \sum_{i \in S^L} \delta_i((\sum_{j \in S} \exp(\Lambda_{a_{ij}(c)})) - 1)$ with Lagrange multiplier δ_i. In analogy, one obtains for each state context $i = (i_1, \ldots, i_L) \in S^L$ and each next state $j \in S$ the transition probability

$$a_{ij}(c)^{(h+1)} = \frac{\left(\sum_{k=1}^{K} \sum_{\substack{t=L \\ c_t(k)=c}}^{T_k-1} \varepsilon_t^k(i,j)\right) + \vartheta_{ij}(c)}{\left(\sum_{v \in S} \sum_{k=1}^{K} \sum_{\substack{t=L \\ c_t(k)=c}}^{T_k-1} \varepsilon_t^k(i,v)\right) + \left(\sum_{v \in S} \vartheta_{iv}(c)\right)} \qquad (3.37)$$

for a transition from the current state i_L to the next state j in transition class $c \in C$ in consideration of the memory $(i_1, \ldots, i_{L-1})$ of the current state. The required Epsilon-Variables $\varepsilon_t^k(i,j)$ and $\varepsilon_t^k(i,v)$ defined in (3.20) have to be computed under the current inhomogeneous $HHMM(L,C)$ $\lambda(h)$. Again, the transition probability $a_{ij}(c)^{(h+1)}$ in (3.37) obtained for the Bayesian Baum-Welch algorithm is extended to that given in (3.26) for the Baum-Welch algorithm by additionally including the prior parameter $\vartheta_{ij}(c)$ to express prior knowledge for this transition probability.

Emission Parameters

Each state $i \in S$ of the inhomogeneous $HHMM(L,C)$ has a state-specific Gaussian emission density $b_i(o)$ defined in (3.2). Each Gaussian emission density is characterized by the state-specific mean μ_i and by the state-specific standard deviation σ_i. To estimate the mean $\mu_i^{(h+1)}$ and the standard deviation $\sigma_i^{(h+1)}$ of the next inhomogeneous $HHMM(L,C)$ $\lambda(h+1)$ one has to maximize the sum of Baum's auxiliary function $Q_3(B \mid \lambda(h))$ given in (3.23) in combination with the log-prior $log(D_3(B \mid \Theta_3))$ defined in (3.32). This is done by computing the critical points of the derivatives of $Q_3(B \mid \lambda(h)) + \log(D_3(B \mid \Theta_3))$ with respect to the mean μ_i and the standard deviation

3. Hidden Markov Models

σ_i. Based on this, the mean of state $i \in S$ for the next inhomogeneous $HHMM(L,C)$ $\lambda(h+1)$ is given by

$$\mu_i^{(h+1)} = \frac{\left(\sum_{k=1}^{K}\sum_{t=1}^{T_k} o_t(k) \cdot \gamma_t^k(i)\right) + \epsilon_i \eta_i}{\left(\sum_{k=1}^{K}\sum_{t=1}^{T_k} \gamma_t^k(i)\right) + \epsilon_i} \quad (3.38)$$

with respect to the state posterior $\gamma_t^k(i)$ given in (3.6) computed under the current inhomogeneous $HHMM(L,C)$ $\lambda(h)$. The mean $\mu_i^{(h+1)}$ in (3.38) obtained in the context of the Bayesian Baum-Welch algorithm is very similar to that given for the Baum-Welch algorithm in (3.27). In comparison to (3.27), the scale parameter ϵ_i allows to quantify the influence of the assumed a priori mean η_i on the mean $\mu_i^{(h+1)}$ in (3.38). The standard deviation of state $i \in S$ of the next inhomogeneous $HHMM(L,C)$ $\lambda(h+1)$ is specified by

$$\sigma_i^{(h+1)} = \sqrt{\frac{\left(\sum_{k=1}^{K}\sum_{t=1}^{T_k} \left(o_t(k) - \mu_i^{(h+1)}\right)^2 \cdot \gamma_t^k(i)\right) + \epsilon_i(\mu_i^{(h+1)} - \eta_i)^2 + 2\alpha_i}{\left(\sum_{k=1}^{K}\sum_{t=1}^{T_k} \gamma_t^k(i)\right) + 2r_i + 2}} \quad (3.39)$$

under consideration of the state posterior $\gamma_t^k(i)$ in (3.6) that is computed under the current inhomogeneous $HHMM(L,C)$ $\lambda(h)$. Again, the standard deviation $\sigma_i^{(h+1)}$ in (3.39) obtained for the Bayesian Baum-Welch algorithm is similar to that given for the Baum-Welch algorithm in (3.28). In addition to (3.28), the quadratic difference of the mean $\mu_i^{(h+1)}$ and the a priori mean η_i is quantified by the scale parameter ϵ_i, and the scale parameter α_i and the shape parameter r_i allow to adjust $\sigma_i^{(h+1)}$ in (3.39). For further reading to the Bayesian estimation of the emission parameters one can consider the article by Lee et al. (1990) and the series of articles by Gauvain and Lee (1991), Gauvain and Lee (1992), and by Gauvain and Lee (1994).

3.7.3 Computational Scheme of the Bayesian Baum-Welch Algorithm

In analogy to the Baum-Welch algorithm, the computational scheme of the Bayesian Baum-Welch algorithm is specified in terms of an initialization and an iteration step.

- *Initialization*: Choose initial state probabilities, transition probabilities, and emission parameters of the inhomogeneous $HHMM(L,C)$ $\lambda(1)$.

- *Iteration*: For iteration steps $h = 1, 2, \ldots$
    - Use the current inhomogeneous $HHMM(L,C)$ $\lambda(h)$ to compute all State-Posterior-Variables $\gamma_t^k(i)$ given in (3.6) and all Epsilon-Variables $\varepsilon_t^k(i,j)$ given in (3.19) and (3.20) based on the given emission sequences $\vec{o}(1), \ldots, \vec{o}(K)$ and their corresponding transition class sequences $\vec{c}(1), \ldots, \vec{c}(K)$.
    - Compute the optimal parameters of the next inhomogeneous $HHMM(L,C)$ $\lambda(h+1)$ based on the previous computations.
        1. Compute the initial state probability $\pi_i^{(h+1)}$ given in (3.35) for each state $i \in S$.
        2. Compute the transition probability $a_{ij}(c)^{(h+1)}$ for each state context $i \in S^t$ of length $1 \le t < L$ and each next state $j \in S$ using (3.36). In analogy, use (3.37) to compute the transition probability for each state context $i \in S^L$ of length L.
        3. Compute the mean $\mu_i^{(h+1)}$ and the standard deviation $\sigma_i^{(h+1)}$ for each state $i \in S$ as specified in (3.38) and (3.39).
    - Stop if the log-posterior of the next inhomogeneous $HHMM(L,C)$ $\lambda(h+1)$ has increased less than a pre-defined threshold in comparison to the log-posterior of the current inhomogeneous $HHMM(L,C)$ $\lambda(h)$, otherwise start the next iteration step with $h := h + 1$.

4 Parsimonious Higher-Order Hidden Markov Models

The inhomogeneous $HHMM(L,C)$ is based on its internal inhomogeneous MM of order L with C transition classes. The basic characteristic of this $HHMM(L,C)$ is that a transition from the current state to the next state is depending on the memory of the $L-1$ predecessor states of the current state. A drawback of the integration of such a memory is the huge increase of the number of transition parameters of the inhomogeneous $HHMM(L,C)$ for an increasing order L. To overcome this, the initial work on homogeneous parsimonious higher-order MMs by Bourguignon and Robelin (2004) and further extensions by Gohr (2006) provide the basis for the development of the inhomogeneous Parsimonious Higher-order Hidden Markov Model of order L with C transition classes ($PHHMM(L,C)$). The general idea behind this $PHHMM(L,C)$ is to reduce the number of transition parameters by introducing equivalence classes of transition parameters. That means, depending on the given data set, selected transition parameters which have been grouped together into one equivalence class have to share common transition probabilities. This sharing of transition parameters reduces the total number of free transition parameters. The equivalence classes of transition parameters are computed efficiently by a dynamic programming algorithm described by Bourguignon and Robelin (2004) and Gohr (2006). Here, it is referred to this algorithm as the Parsimonious Cluster algorithm. This algorithm is adapted to the specific requirements of the $PHHMM(L,C)$ through the integration into the framework of the Bayesian Baum-Welch algorithm.

Goals of this Chapter

1. Partitions of the set of hidden states S are introduced to provide the basics for equivalence classes of transition parameters.

2. A tree-based representation of state contexts over S is described to represent the

4. Parsimonious Higher-Order Hidden Markov Models

equivalence classes of transition parameters.

3. The definition of the inhomogeneous *PHHMM*(L, C) is given.

4. In the context of the *Maximum A Posteriori Problem* the Bayesian Baum-Welch algorithm is extended to the inhomogeneous *PHHMM*(L, C) by integrating the Parsimonious Cluster algorithm to determine the optimal equivalence classes of transition parameters.

4.1 Partitions of the Set of Hidden States

The basic idea behind a partition is to group certain states of the set of hidden states S into disjoint subsets of equivalent states. The partitions of S are used to define parsimonious representations of state contexts over S. These parsimonious representations are considered in the next section. Here, the focus is to establish the theoretical basics behind.

4.1.1 Computing the Partitions

First, the definition of a partition is required to compute all partitions of the set of hidden states S. A partition ρ of the set of hidden states S is defined by the following four properties.

1. The partition ρ is subset of the power set of S.
2. The partition ρ does not contain the empty set.
3. All elements of the partition ρ are mutually disjoint.
4. The union set of all elements of the partition ρ is identical to S.

These properties ensure that each state $i \in S$ is represented by exactly one element of a partition. Based on this definition, one can derive an algorithm that iteratively computes the set of all partitions $\Delta(N)$ for the set of N hidden states $S := \{S_1, \ldots, S_N\}$.

- *Initialization*: The set of partitions of $S = \{S_1\}$ is $\Delta(1) = \{\{\{S_1\}\}\}$.

- *Iteration*: For each iteration step $n = 1, 2, \ldots, N-1$ and its corresponding set of partitions $\Delta(n)$.

1. Set $\Delta(n+1) := \{\}$.
2. Add new partitions to $\Delta(n+1)$ by extending $\Delta(n)$ with the new state S_{n+1} using the two following rules on each partition $\rho = \{\rho_1, \ldots, \rho_k\} \in \Delta(n)$ of any cardinality $k \geq 1$.
   - R1: Add $\{\rho_1, \ldots, \rho_k, \{S_{n+1}\}\}$ to $\Delta(n+1)$.
   - R2: Add $\{\rho_1, \ldots, \rho_{h-1}, \rho_h \cup \{S_{n+1}\}, \rho_{h+1}, \ldots, \rho_k\}$ to $\Delta(n+1)$ for each set $\rho_h \in \rho$ with index $1 \leq h \leq k$.

This algorithm computes all partitions $\Delta(n+1)$ of the set $S := \{S_1, \ldots, S_{n+1}\}$ of $n+1$ hidden states by extending each partition of $\Delta(n)$ by the new state S_{n+1}. The extensions are done with respect to the definition of the partition. That is, rule R1 extends each partition $\rho \in \Delta(n)$ by the new element $\{S_{n+1}\}$. This leads to a well-defined partition of $n+1$ hidden states, because each state S_i with $1 \leq i \leq n$ is already contained in exactly one element of the considered partition ρ. The rule R2 adds all partitions to $\Delta(n+1)$ that result from the combination of the new state S_{n+1} with each specific element of a partition $\rho \in \Delta(n)$.

4.1.2 Number of Partitions

The number of partitions in $\Delta(n)$ for a set $S := \{S_1, \ldots, S_n\}$ of n hidden states is given by the Bell number

$$B[n] := \sum_{k=1}^{n} S[n, k] \qquad (4.1)$$

like defined by Pitman (1997). The Bell number can be computed by summing over the Stirling number of the second kind $S[n, k] := S[n-1, k-1] + k \cdot S[n-1, k]$, which is defined in Abramowitz and Stegun (1972). The Stirling number represents the number of different ways to partition a set of n elements into exactly k non-empty subsets. The computation of the Stirling number is done with respect to the initial value $S[1, 1] = 1$, and in consideration of the constraints $S[n, 0] = 0$ for all $n \in \mathbb{N}$, and $S[n, k] = 0$ for all $k \in \mathbb{N}$ greater than $n \in \mathbb{N}$. The first term $S[n-1, k-1]$ on the right-hand side of $S[n, k]$ represents the number of partitions that are obtained by applying the rule R1 to each partition $\rho \in \Delta(n-1)$ of cardinality $k-1$, and the second term $k \cdot S[n-1, k]$ counts the number of partitions that result from the application of rule R2 to all partitions $\rho \in \Delta(n-1)$ of cardinality k.

4. Parsimonious Higher-Order Hidden Markov Models

4.1.3 Set of Partitions

The set of partitions of the set of N hidden states S contains Bell number $B[N]$ partitions and is defined by

$$\Delta_S := \Delta(N) \tag{4.2}$$

for the inhomogeneous *PHHMM*(L,C). Subsequently, Δ_S is used to define the tree-based representation of state contexts over S.

4.2 Tree-based Representation of State Contexts

The basis of the inhomogeneous *PHHMM*(L,C) is its internal inhomogeneous higher-order *MM* specified in Sec. 2.2.2. Here, a transition from the current state $i_l \in S$ to a next state $j \in S$ is depending on the memory $(i_1, \ldots, i_{l-1})$ of the current state i_l. Each individual state context $(i_1, \ldots, i_l) \in S^l$ and its corresponding transition parameters can be represented by a tree-based data structure using the set of partitions Δ_S defined in (4.2). To develop this data structure, the height of a tree is defined to be the number of edges on the path from the root node to the deepest node in the tree. In analogy to this, the depth of a node in a tree is defined to be the number of edges on the path from the root node to the considered node. Based on this, the tree τ_l of height l that stores all state contexts of length l has to fulfill the following properties.

1. The root node n in depth 0 is labeled by the set $\mathcal{L}[n] := \{\epsilon\}$ containing the empty word ϵ.

2. Each node v in depth $d_v \in \{1, \ldots, l\}$ is linked to its parent node $\mathcal{P}[v]$ in depth $d_v - 1$, and each v is labeled by a non-empty subset $\mathcal{L}[v]$ of the set of hidden states S.

3. The set of labels of all child nodes of each parent node defines a partition in the set of partitions Δ_S.

4. All leaf nodes are in depth l.

5. Each leaf v defines a set of equivalent state contexts $\xi[v,l] := \{(i_1, i_2, \ldots, i_l) : i_1 \in \mathcal{L}[v], i_2 \in \mathcal{L}[\mathcal{P}[v]], \ldots, \epsilon \in \mathcal{L}[n]\}$ of length l. The state contexts of leaf v represent all combinations of states that are obtained by traversing the path from the leaf node v to the root node n.

4. Parsimonious Higher-Order Hidden Markov Models

This tree ensures that each state of S is included exactly in one label of the child nodes of each parent node. That is because the set of labels of all child nodes of a parent node is defined to be a partition of S. This implicates that each state context $(i_1, \ldots, i_l) \in S^l$ is contained exactly in one of the sets of equivalent state contexts that are represented by the leaf nodes of the tree. Based on this, each tree τ_l defines a set of equivalence classes of state contexts

$$\xi_{\tau_l} := \{\xi[v,l] : v \text{ is a leaf node of the tree } \tau_l\} \qquad (4.3)$$

that partitions all different state contexts $(i_1, \ldots, i_l) \in S^l$ into sets of equivalent state contexts. The number of different equivalence classes defines three types of trees. The completely fused tree that represents all state contexts in one equivalence class, the complete tree that represents each state context in a separate equivalence class, and the parsimonious tree that groups individual state contexts together resulting in less equivalence classes than in a complete tree. The complete tree is associated with the transition parameters of the inhomogeneous $HHMM(L,C)$, and the completely fused or the parsimonious tree represent the transition parameters of the inhomogeneous $PHHMM(L,C)$. Due to the fact that the state contexts in an equivalence class of ξ_{τ_l} are defined to be equivalent, the individual transition parameters of these state contexts are replaced by transition parameters that are shared by all state contexts of this equivalence class.

To provide an overview, all different trees of height one and three selected parsimonious trees of height two are shown in Fig. 4.1 with respect to the set of hidden states $S := \{-, =, +\}$. Based on (4.1), S has five different partitions that are given by $\Delta_S = \{\{\{-, =, +\}\}, \{\{-, =\}, \{+\}\}, \{\{-, +\}, \{=\}\}, \{\{-\}, \{=, +\}\}, \{\{-\}, \{+\}, \{=\}\}\}$ with respect to (4.2). Each of these different partitions is represented by an individual tree of height one in Fig. 4.1. Each leaf node of such a tree defines an equivalence class of state contexts. For instance, the completely fused tree of height one represents all state contexts $i \in S^1$ of length one by the equivalence class $\{(-), (=), (+)\}$, while the first one of the three parsimonious trees represents these state contexts by the two equivalence classes $\{(-), (=)\}$ and $\{(+)\}$. Three different parsimonious trees of height two are shown in the bottom part of Fig. 4.1. These trees are obtained by extending the leaf nodes of the corresponding tree of height one with the particular partitions of S. For example, the first tree represents all state contexts $i \in S^2$ of length two by the two equivalence classes $\{(-,-), (-,=), (-,+)\}$ and $\{(=,-), (=,=), (=,+), (+,-), (+,=), (+,+)\}$, while the second tree has the two equivalence classes

4. Parsimonious Higher-Order Hidden Markov Models

$\{(-,-),(-,+),(=,-),(=,+),(+,-),(+,+)\}$ and $\{(-,=),(=,=),(+,=)\}$ to represent these state contexts. The three parsimonious trees of height two shown in Fig. 4.1 are selected from the 205 different trees of height two that can be constructed based on the five partitions of S. How the number of different trees develops for increasing tree height is shown exemplarily in Fig. 4.2 for S containing two and three states. The huge increase of the number of different trees for an increasing tree height requires an efficient strategy that is capable to evaluate these trees. This strategy is subsequently developed for the $PHHMM(L, C)$ in the following sections by integrating the Parsimonious Cluster algorithm developed by Bourguignon and Robelin (2004) and Gohr (2006).

All different trees of height 1

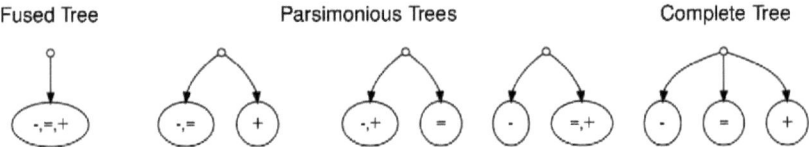

Selected parsimonious trees of height 2

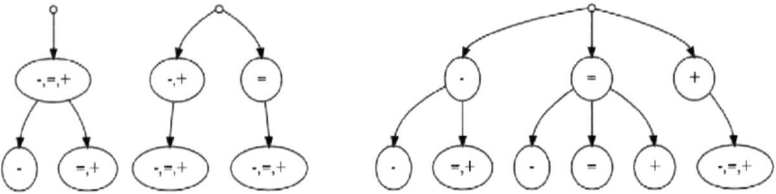

Figure 4.1: Overview of different types of trees of height one and two based on the set of hidden states $S := \{-, =, +\}$. Each individual tree τ_1 of height one that can be constructed from the five partitions of S is shown in the top part. Three selected parsimonious trees τ_2 of height two out of the 205 different trees of height two that can be generated based on the five partitions of S are shown in the bottom part. The increase of the number of trees in dependency of the tree height is shown in Fig. 4.2.

4. Parsimonious Higher-Order Hidden Markov Models

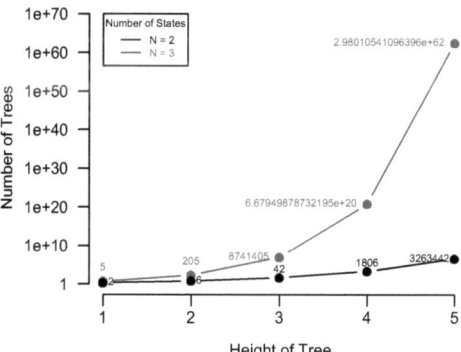

Figure 4.2: Overview of the number of different trees that exist for $N = 2$ or $N = 3$ hidden states $S := \{S_1, \ldots, S_N\}$ for increasing tree height. The number of trees are shown in logarithmic scale. Individual numbers are shown explicitly within the plot. The number of trees has been computed in analogy to the computational scheme of the Parsimonious Cluster algorithm.

4.3 Inhomogeneous Parsimonious Higher-Order Hidden Markov Model

The inhomogeneous *PHHMM*(L, C) of order L with C different transition classes is defined by $\lambda = (\vec{\pi}, \mathcal{T}, A, B)$ with respect to its following parameters.

1. The initial state distribution $\vec{\pi} := (\pi_{S_1}, \ldots, \pi_{S_N})$ defines for each state $i \in S$ the probability $\pi_i := P[Q_1 = i]$ of starting in this state at time step $t = 1$. Two stochastic constraints must be fulfilled by $\vec{\pi}$.

    a) $\forall i \in S : \pi_i \in [0, 1]$

    b) $\sum_{i \in S} \pi_i = 1$

2. The set of tree structures $\mathcal{T} := \{\mathcal{T}_1, \ldots, \mathcal{T}_C\}$ defines for each transition class $c \in \mathcal{C}$ the trees $\mathcal{T}_c := (\tau_1(c), \ldots, \tau_L(c))$. Each tree $\tau_l(c)$ represents the equivalence classes $\xi_{\tau_l(c)}$ of state contexts of length l. The equivalence classes are defined in (4.3) based on the tree-based representation of state contexts.

3. The set of transition matrices $A := \{A_1, \ldots, A_C\}$ defines for each transition class $c \in \mathcal{C}$ the transition matrix $A_c := (a_{ij}(c))$ with respect to the given trees $\mathcal{T}_c$. That

57

4. Parsimonious Higher-Order Hidden Markov Models

is, each A_c defines for each state context $i = (i_1, \ldots, i_l) \in S^l$ of length $1 \leq l \leq L$ and each next state $j \in S$ the transition probability

$$a_{ij}(c) := \begin{cases} P[Q_{t+1} = j \mid \vec{Q}_{1\ldots t} = i, c], & 1 \leq t < L \\ P[Q_{t+1} = j \mid \vec{Q}_{t-L+1\ldots t} = i, c], & t \geq L \end{cases}$$

for a transition from the current state i_l to the next state j at time step t using transition class c with respect to the memory $(i_1, \ldots, i_{l-1})$ of the current state. Additionally, all state contexts i that are contained in an equivalence class $\xi \in \xi_{\tau_l(c)}$ given by the tree $\tau_l(c)$ are defined to have the identical transition probabilities $a_{\xi j}(c_t)$. That is, all state contexts $i \in \xi$ share their transition probabilities. Again, the transition probabilities of each equivalence class ξ have to fulfill two stochastic constraints.

a) $\forall j \in S : a_{\xi j}(c) \in [0, 1]$

b) $\sum_{j \in S} a_{\xi j}(c) = 1$

4. The matrix $B := (\mu_i, \sigma_i)$ defines the state-specific mean $\mu_i \in \mathbb{R}$ and the state-specific standard deviation $\sigma_i \in \mathbb{R}^+$ for the Gaussian emission density of each state $i \in S$. The time-independent probability density $b_i(o) := P[O_t = o \mid Q_t = i]$ for emitting an emission $o \in \mathbb{R}$ by the Gaussian emission density of state i is defined in (3.2).

The inhomogeneous *PHHMM(L, C)* reduces to the homogeneous *PHHMM(L)* for $C = 1$ transition class. With respect to the notation scheme in Tab. 3.1, the inhomogeneous *PHHMM(L, C)* with underlying complete trees represents the inhomogeneous *HHMM(L, C)*. Thus, the inhomogeneous *PHHMM(L, C)* also includes the homogeneous *HMM*, the inhomogeneous *HMM(C)*, and the homogeneous *HHMM(L)* as special cases.

4.4 Solving the Maximum A Posteriori Problem

The *Maximum A Posteriori Problem* has been solved for the inhomogeneous *HHMM(L, C)* based on the Bayesian Baum-Welch algorithm that iteratively optimizes Baum's auxiliary function $Q(\lambda \mid \lambda(h))$ given in (3.10) in combination with the prior $P[\lambda]$ specified in (3.29). For the inhomogeneous *PHHMM(L, C)* one has to adapt Baum's auxiliary function and the prior by taking into account that the set of transition matrices

A is now depending on the set of tree structures $\mathcal{T}$. For that reason, the two classes of Baum's auxiliary functions $Q_2^t(A \mid \lambda(h))$ for transition probabilities in (3.21) and (3.22) have to be modified in consideration of the tree structures $\mathcal{T}$. In addition to this, the transition prior $D_2^t(A \mid \Theta_2)$ given in (3.31) has to be modified to represent the dependencies introduced by the set of tree structures $\mathcal{T}$. Finally, a prior for scoring all different tree structures must be considered. Subsequently, the focus is on realizing these adaptations to establish the basics for the extension of the Bayesian Baum-Welch algorithm by integrating the Parsimonious Cluster algorithm. Here, the Parsimonious Cluster algorithm is the key component for the computation of the optimal transition parameters of the *PHHMM(L, C)*.

4.4.1 Tree-Based Baum's Auxiliary Function for Transition Parameters

For the transition matrix A_c all state contexts $i \in S^t$ of length t are represented by the corresponding tree $\tau_t(c)$. Each $\tau_t(c)$ represents a set of transition class specific equivalence classes $\xi_{\tau_t(c)}$ of state contexts like generally defined in (4.3). All state contexts i that are contained in an equivalence class $\xi \in \xi_{\tau_t(c)}$ are defined to have an identical transition probability $a_{\xi j}(c)$ for each transition to a next state $j \in S$ using the transition class $c \in \mathcal{C}$. These constraints must be integrated into Baum's auxiliary function for transition parameters of the *HHMM(L, C)* to obtain the corresponding functions for the *PHHMM(L, C)*. Thus, Baum's auxiliary function $Q_2^t(A \mid \lambda(h))$ in (3.21) for transition parameters used by the inhomogeneous *HHMM(L, C)* at time steps $1 \leq t < L$ is modified to

$$Q_2^t(A \mid \lambda(h)) := \sum_{c \in \mathcal{C}} \sum_{\xi \in \xi_{\tau_t(c)}} \sum_{j \in S} \Lambda_{a_{\xi j}(c)} \sum_{\substack{k=1 \\ c_t(k)=c}}^{K} \sum_{i \in \xi} \varepsilon_t^k(i,j) \qquad (4.4)$$

for estimating the corresponding transition probabilities of the next inhomogeneous *PHHMM(L, C)* $\lambda(h+1)$. Here, the sum over all individual state contexts $i \in S^t$ in (3.21) is replaced by the sum over all equivalence classes $\xi \in \xi_{\tau_t(c)}$ of state contexts represented by the corresponding tree $\tau_t(c)$. This leads to the substitution of the individual logarithmic transition probability $\Lambda_{a_{ij}(c)} := \log(a_{ij}(c))$ in (3.21) by the corresponding parameter $\Lambda_{a_{\xi j}(c)} := \log(a_{\xi j}(c))$. Due to that, an additional sum over all state contexts $i \in \xi$ has to be included for adding up all individual Epsilon-Variables $\varepsilon_t^k(i,j)$ in (3.21). In analogy, the same modifications are made for Baum's auxiliary function $Q_2^L(A \mid \lambda(h))$

4. Parsimonious Higher-Order Hidden Markov Models

in (3.22) leading to

$$Q_2^L(A\,|\,\lambda(h)) := \sum_{c \in \mathcal{C}} \sum_{\xi \in \xi_{\tau_L(c)}} \sum_{j \in S} \Lambda_{a_{\xi j}(c)} \sum_{k=1}^{K} \sum_{\substack{t=L \\ \mathbf{c_t(k) = c}}}^{T_k-1} \sum_{i \in \xi} \varepsilon_t^k(i,j) \tag{4.5}$$

for estimating the transition parameters of the next inhomogeneous $PHHMM(L,C)$ $\lambda(h+1)$ used at all time steps $t \geq L$.

4.4.2 Tree-Based Transition Prior

The prior of the transition parameters $D_2^t(A\,|\,\Theta_2)$ of the inhomogeneous $HHMM(L,C)$ is defined to be a product of independent Dirichlet distributions given in (3.31). This prior needs to be adapted for the inhomogeneous $PHHMM(L,C)$. This is done by accounting for the given set of tree structures $\mathcal{T}$. Each tree $\tau_L(c)$ in $\mathcal{T}$ defines a set of equivalence classes $\xi_{\tau_t(c)}$ of state contexts for which all state contexts of an equivalence class $\xi \in \xi_{\tau_t(c)}$ have identical transition probabilities $a_{\xi j}(c)$. This leads to the adapted transition prior for state contexts of length $1 \leq t \leq L$ given by

$$D_2^t(A\,|\,\mathcal{T},\Theta_2) := \prod_{c \in \mathcal{C}} \prod_{\xi \in \xi_{\tau_t(c)}} Z(\Theta_2^\xi(c)) \prod_{j \in S} \exp\left(\Lambda_{a_{\xi j}(c)} \cdot \vartheta_{\xi j}(c)\right) \tag{4.6}$$

in consideration of the set of tree structures $\mathcal{T}$ and the matrices Θ_2 of transition prior parameter $\vartheta_{ij}(c) \in \mathbb{R}^+$ defined for the transition prior in (3.31). The vector $\Theta_2^\xi(c) := (\vartheta_{\xi S_1}(c), \ldots, \vartheta_{\xi S_N}(c))$ with $\vartheta_{\xi j}(c) := \sum_{i \in \xi} \vartheta_{ij}(c)$ defines the transition prior parameters for the state contexts of length t in the equivalence class $\xi \in \xi_{\tau_t(c)}$ of the tree $\tau_t(c)$ of transition class c in $\mathcal{T}$. The normalization constant $Z(\Theta_2^\xi(c)) := \Gamma(\sum_{j \in S} \vartheta_{\xi j}(c))/\prod_{j \in S} \Gamma(\vartheta_{\xi j}(c))$ is based on the Gamma function $\Gamma(x)$ defined for the transition prior in (3.31).

4.4.3 Tree Structure Prior

The tree structure prior must provide the opportunity to differentiate between different realizations of tree structures in $\mathcal{T}$. One way of doing this is to quantify the number of equivalence classes that are defined by a tree. This is realized by using the tree structure prior

$$D_4^t(\mathcal{T}\,|\,\varphi) \propto \prod_{c\in\mathcal{C}} \prod_{\xi\in\xi_{\tau_t(c)}} \varphi \qquad (4.7)$$

for scoring each equivalence class ξ of the tree $\tau_t(c)$ in the set of tree structures $\mathcal{T}$ by the same parameter $\varphi \in \mathbb{R}^+$. That is, for $\varphi \in (0,1)$ the value of the tree structure prior increases if the total number of equivalence classes is decreased, and for $\varphi > 1$ the value of the tree structure prior increases if the total number of equivalence classes is increased. The value of the tree structure prior is independent of the total number of equivalence classes for $\varphi = 1$.

4.4.4 Bayesian Baum-Welch Algorithm

The Bayesian version of the Baum-Welch algorithm represents an iterative training procedure that locally maximizes the posterior of the inhomogeneous $PHHMM(L,C)$ based on the adapted Baum's auxiliary function in combination with the adapted prior. In analogy to the Bayesian Baum-Welch algorithm for the inhomogeneous $HHMM(L,C)$, the computational scheme of this algorithm is specified in detail for the inhomogeneous $PHHMM(L,C)$ in terms of an initialization and an iteration step.

- *Initialization*: Choose initial state probabilities, transition probabilities according to complete trees, and emission parameters for the initial inhomogeneous $PHHMM(L,C)$ $\lambda(1)$.

- *Iteration*: For iteration steps $h = 1, 2, \ldots$
  - Use the current inhomogeneous $PHHMM(L,C)$ $\lambda(h)$ to compute all State-Posterior-Variables $\gamma_t^k(i)$ given in (3.6) and all Epsilon-Variables $\varepsilon_t^k(i,j)$ given in (3.19) and (3.20) based on the given emission sequences $\vec{o}(1), \ldots, \vec{o}(K)$ and their corresponding transition class sequences $\vec{c}(1), \ldots, \vec{c}(K)$.
  - Compute the optimal parameters of the next inhomogeneous $PHHMM(L,C)$ $\lambda(h+1)$ based on the previous computations.
    1. Compute all initial state probabilities $\pi_i^{(h+1)}$ as shown in (3.35) to maximize $Q_1(\vec{\pi}\,|\,\lambda(h)) + \log(D_1(\vec{\pi}\,|\,\Theta_1))$.
    2. Compute the tree structures $\mathcal{T}^{h+1}$ and the corresponding transition parameters $a_{\xi j}(c)^{(h+1)}$ used at time steps $t \geq 1$ by maximizing $Q_2^t(A\,|\,\lambda(h)) + \log(D_2^t(A\,|\,\mathcal{T},\Theta_2)) + \log(D_4^t(\mathcal{T}\,|\,\varphi))$.

4. Parsimonious Higher-Order Hidden Markov Models

3. Compute all state-specific means $\mu_i^{(h+1)}$ and standard deviations $\sigma_i^{(h+1)}$ as shown in (3.38) and (3.39) to maximize $Q_3(B\,|\,\lambda(h)) + \log(D_3(B\,|\,\Theta_3))$.

– Stop if the log-posterior under the next inhomogeneous *PHHMM*(L,C) $\lambda(h+1)$ has increased less than a pre-defined threshold in comparison to the log-posterior under the current inhomogeneous *PHHMM*(L,C) $\lambda(h)$, otherwise start the next iteration step with $h := h+1$.

This computational scheme of the Bayesian Baum-Welch algorithm allows to solve the *Maximum A Posteriori Problem* for the inhomogeneous *PHHMM*(L,C) by starting with initial model parameters that are iteratively adapted until the posterior of this *PHHMM*(L,C) reaches a local optimum. Regarding the iteration step, the initial state probabilities and the state-specific emission parameters are computed as described for the inhomogeneous *HHMM*(L,C). What remains is to find a solution for the computation of the tree structures and their corresponding transition probabilities. That is, the tree structures in T^{h+1} and their corresponding transition parameters $a_{\xi j}(c)^{(h+1)}$ have to be determined by maximizing

$$F_t(A, \mathcal{T}) := Q_2^t(A\,|\,\lambda(h)) + \log(D_2^t(A\,|\,\mathcal{T}, \Theta_2)) + \log(D_4^t(\mathcal{T}\,|\,\varphi)) \qquad (4.8)$$

based on Baum's auxiliary functions for transition parameters $Q_2^t(A\,|\,\lambda(h))$ given in (4.4) for $1 \leq t < L$ and given in (4.5) for $t \geq L$ in consideration of the transition prior $D_2^t(A\,|\,\mathcal{T}, \Theta_2)$ in (4.6) and the tree structure prior $D_4^t(\mathcal{T}\,|\,\varphi)$ in (4.7). This is done subsequently by introducing a scoring scheme for the different tree structures that allows to compute the optimal set of tree structures and their corresponding optimal transition probabilities.

4.4.5 Scoring Scheme for Tree Structures

The objective function $F_t(A, \mathcal{T})$ defined in (4.8) for the computation of the optimal tree structures and their corresponding transition probabilities can be alternatively expressed by

$$F_t(A, \mathcal{T}) = \sum_{c \in \mathcal{C}} \sum_{\xi \in \xi_{\tau_t(c)}} f_t(\vec{a}_\xi(c)) \qquad (4.9)$$

4. Parsimonious Higher-Order Hidden Markov Models

in terms of a score function $f_t(\vec{a}_\xi(c))$ that evaluates each equivalence class ξ of state contexts of length t in transition class c based on the corresponding transition probabilities $\vec{a}_\xi(c) := (a_{\xi S_1}(c), \ldots, a_{\xi S_N}(c))$. By regrouping and conflating of the individual terms in (4.8), the score function

$$f_t(\vec{a}_\xi(c)) := h_t(\vec{a}_\xi(c)) + \log(\varphi) + \log\left(Z(\Theta_2^\xi(c))\right) \qquad (4.10)$$

is obtained. This function consists of a function $h_t(\vec{a}_\xi(c))$, the value $\log(\varphi)$ for scoring the equivalence class, and the value $\log(Z(\Theta_2^\xi(c)))$ of the normalization constant of the transition prior. This score function can be used to determine the score of any given equivalence class ξ. The score of each equivalence class ξ can be maximized by estimating the corresponding transition probabilities $\vec{a}_\xi(c)^{(*)}$. For the transition parameters that are used at a fixed time step $1 \leq t < L$, this is done by maximizing the function

$$h_t(\vec{a}_\xi(c)) := \sum_{j \in S} \Lambda_{a_{\xi j}(c)} \left(\left(\sum_{\substack{k=1 \\ c_t(k)=c}}^{K} \sum_{i \in \xi} \varepsilon_t^k(i,j) \right) + \vartheta_{\xi j}(c) \right) \qquad (4.11)$$

based on all state contexts i of a length t that are contained in ξ. Here, the Epsilon-Variable $\varepsilon_t^k(i,j)$ defined in (3.19) is computed under the current inhomogeneous $PHHMM(L,C)$ $\lambda(h)$ in the iteration step of the Bayesian Baum-Welch algorithm, and $\vartheta_{\xi j}(c)$ represents the transition prior constant for the corresponding transition. For time steps $t \geq L$, the function $h_L(\vec{a}_\xi(c))$ is defined by

$$h_L(\vec{a}_\xi(c)) := \sum_{j \in S} \Lambda_{a_{\xi j}(c)} \left(\left(\sum_{k=1}^{K} \sum_{\substack{t=L \\ c_t(k)=c}}^{T_k-1} \sum_{i \in \xi} \varepsilon_t^k(i,j) \right) + \vartheta_{\xi j}(c) \right) \qquad (4.12)$$

in consideration of all state contexts i of length L represented by the equivalence class ξ and the corresponding Epsilon-Variable $\varepsilon_t^k(i,j)$ defined in (3.20) computed under the current inhomogeneous $PHHMM(L,C)$ $\lambda(h)$. Subsequently, the focus is on the general estimation of transition probabilities for an equivalence class ξ to maximize $f_t(\vec{a}_\xi(c))$ in (4.10). Then, the Parsimonious Cluster algorithm is considered to efficiently compute the optimal tree structure involving the evaluation of all possible equivalence classes and their corresponding optimal scores. Taking this together, this algorithm determines the optimal tree structures T^{h+1} and their corresponding transition parameters $a_{\xi j}(c)^{(h+1)}$ to maximize $F_t(A, T)$ in (4.9) for each state context length $1 \leq t \leq L$.

4. Parsimonious Higher-Order Hidden Markov Models

4.4.6 Estimating Transition Parameters for an Equivalence Class

An equivalence class ξ of a tree $\tau_t(c)$ represents state contexts $i \in S^t$ that are defined to be equivalent by having an identical transition probability $a_{\xi j}(c)$ for a transition to each state $j \in S$. The basis for the estimation of these transition probabilities is the function $h_t(\vec{a}_\xi(c))$ defined in (4.11) for state contexts of length $1 \leq t < L$ and given in (4.12) for state contexts of length L. This function is maximized in subject to the constraint $\sum_{j \in S} \exp(\Lambda_{a_{\xi j}(c)}) = 1$ using the auxiliary function $h_t(\vec{a}_\xi(c)) - \delta \cdot ((\sum_{j \in S} \exp(\Lambda_{a_{\xi j}(c)}) - 1)$ with Lagrange multiplier δ. The auxiliary function is differentiated with respect to $\Lambda_{a_{\xi j}(c)}$ and δ. Both derivatives are set equal to zero to compute the optimal transition probabilities for the equivalence class ξ. For state contexts of a fixed length $1 \leq t < L$, this leads for each state context $i \in \xi$ to the optimal transition probability

$$a_{\xi j}(c)^{(*)} = \frac{\left(\sum_{\substack{k=1 \\ c_t(k)=c}}^{K} \sum_{i \in \xi} \varepsilon_t^k(i,j) \right) + \vartheta_{\xi j}(c)}{\left(\sum_{i \in \xi} \sum_{v \in S} \sum_{\substack{k=1 \\ c_t(k)=c}}^{K} \varepsilon_t^k(i,v) \right) + \left(\sum_{v \in S} \vartheta_{\xi v}(c) \right)} \quad (4.13)$$

used at the fixed time step t for a transition from the current state of each state context i to the next state $j \in S$ under consideration of the relation $a_{\xi j}(c) = \exp(\Lambda_{a_{\xi j}(c)})$. In analogy to this, each equivalence class ξ that represents state contexts of length L has the corresponding optimal transition probability

$$a_{\xi j}(c)^{(*)} = \frac{\left(\sum_{k=1}^{K} \sum_{\substack{t=L \\ c_t(k)=c}}^{T_k-1} \sum_{i \in \xi} \varepsilon_t^k(i,j) \right) + \vartheta_{\xi j}(c)}{\left(\sum_{i \in \xi} \sum_{v \in S} \sum_{k=1}^{K} \sum_{\substack{t=L \\ c_t(k)=c}}^{T_k-1} \varepsilon_t^k(i,v) \right) + \left(\sum_{v \in S} \vartheta_{\xi v}(c) \right)} \quad (4.14)$$

for a transition from the current state of each state context $i \in \xi$ to the next state $j \in S$ at time steps $t \geq L$. In both cases, the Epsilon-Variables $\varepsilon_t^k(i,j)$ and $\varepsilon_t^k(i,v)$ given in (3.19) and (3.20) are computed in the iteration step of the Bayesian Baum-Welch algorithm using the current inhomogeneous *PHHMM*(L,C) $\lambda(h)$.

The structures of the estimation formulas (4.13) and (4.14) follow those obtained in

(3.36) and (3.37) for the inhomogeneous $HHMM(L,C)$. The differences result form the usage of the equivalence class ξ. That is, an additional sum over all state contexts $i \in \xi$ is used to represent all state context in ξ, and the parameters $\vartheta_{\xi j}(c)$ and $\vartheta_{\xi v}(c)$ of the transition prior have been adapted as specified for (4.6). The transition parameters obtained in (4.13) and (4.14) maximize the function $h_t(\vec{a}_\xi(c))$. This can be proven like outlined by Durbin et al. (1998) for the transition parameters of a homogeneous HMM. In addition to this, also the score function $f_t(\vec{a}_\xi(c))$ in (4.10) is maximized by these transition probabilities, because for a given equivalence class ξ only the term $h_t(\vec{a}_\xi(c))$ is variable while the other two terms are constants. Subsequently, this score function is used to determine the optimal tree structures, their corresponding optimal equivalence classes of state contexts, and their corresponding optimal transition parameters for the next inhomogeneous $PHHMM(L,C)$ $\lambda(h+1)$.

4.4.7 Basics for Determining Optimal Tree Structures and Corresponding Transition Parameters

The tree $\tau_t(c)$ represents each state contexts $i \in S^t$ of length t. The leaves of the tree define as specified in (4.3) the set of equivalence classes $\xi_{\tau_t(c)}$ of these state contexts, and all state contexts that are contained in an equivalence class $\xi \in \xi_{\tau_t(c)}$ are defined to have the identical transition probability $a_{\xi j}(c)$ for each transition to a next state $j \in S$. The optimal score of the tree $\tau_t(c)$ is given by the sum of the optimal scores of its equivalence classes defined by

$$F(\tau_t(c)) := \sum_{\xi \in \xi_{\tau_t(c)}} f_t(\vec{a}_\xi(c)^{(*)}) \qquad (4.15)$$

with respect to the optimal score $f_t(\vec{a}_\xi(c)^{(*)})$ specified in (4.10) for each equivalence class $\xi \in \xi_{\tau_t(c)}$. The optimal score $f_t(\vec{a}_\xi(c)^{(*)})$ is computed using the corresponding optimal transition probabilities $\vec{a}_\xi(c)^{(*)} := (a_{\xi S_1}(c)^{(*)}, \ldots, a_{\xi S_N}(c)^{(*)})$ given in (4.13) for $1 \leq t < L$ or given in (4.14) for $t = L$. Moreover, the optimal score function $F(\tau_t(c))$ in (4.15) is part of the global score function $F_t(A, \mathcal{T})$ in (4.9). Here, $F(\tau_t(c))$ quantifies the individual contribution of the tree $\tau_t(c)$ to the global score function. Thus, $F(\tau_t(c))$ can be used to compute the optimal tree $\tau_t(c)^{(h+1)}$ and its corresponding optimal transition parameters $a_{\xi j}(c)^{(h+1)}$ for the next inhomogeneous $PHHMM(L,C)$ $\lambda(h+1)$. For that reason, an extended tree is specified subsequently. This extended tree allows the efficient evaluation of all different tree structures of $\tau_t(c)$ with the Parsimonious Cluster algorithm by Gohr (2006).

4.4.8 Extended Tree

Each state context $i \in S^t$ of length t must be contained in exactly one equivalence class of state contexts of the tree $\tau_t(c)$. To enable the efficient evaluation of all different equivalence classes of all state contexts of length t, the extended tree ψ_t of height t is defined to have the following properties.

1. The root node n in depth 0 is labeled by the set $\mathcal{L}[n] := \{\epsilon\}$ containing the empty word ϵ.

2. Each node v in depth $d_v \in \{1, \ldots, t\}$ is linked to its parent node $\mathcal{P}[v]$ in depth $d_v - 1$, and each v is labeled by a non-empty subset $\mathcal{L}[v]$ of the set of hidden states S.

3. Each node v in depth $d_v \in \{0, \ldots, t-1\}$ has $2^N - 1$ child nodes with labels $\mathcal{L}[\cdot]$ that represent all non-empty elements of the power set of S.

4. All leaf nodes are in depth t.

5. Each leaf v represents a set of equivalent state contexts $\xi[v,t] := \{(i_1, i_2 \ldots, i_t) : i_1 \in \mathcal{L}[v], i_2 \in \mathcal{L}[\mathcal{P}[v]], \ldots, \epsilon \in \mathcal{L}[n]\}$ of length t. The state contexts of leaf v define all combinations of states that are obtained by traversing the path from the leaf node v to the root node n.

The fact that the child nodes of each non-leaf node represent all different non-empty subsets of the power set of S ensures that each of the different equivalence classes of state contexts of length t is contained in the extended tree ψ_t. Based on this, the set of all equivalence classes is defined by

$$\xi_{\psi_t} := \{\xi[v,t] : v \text{ is a leaf node of the extended tree } \psi_t\} \quad (4.16)$$

in consideration of all equivalence classes given by the extended tree ψ_t. The difference between this extended tree and the tree $\tau_t(c)$ defined in Sec. 4.2 is that $\tau_t(c)$ represents all state contexts of length t by a specific set of disjoint equivalence classes, while the extended tree ψ_t contains all different equivalence classes of these state contexts. Subsequently, the Parsimonious Cluster algorithm by Gohr (2006) is applied to transform the extended tree ψ_t into the optimal tree $\tau_t(c)^{(h+1)}$ for the next inhomogeneous PHHMM(L,C) $\lambda(h+1)$ by selecting the optimal set of disjoint equivalence classes from ξ_{ψ_t} with respect to the tree score function $F(\tau_t(c))$ in (4.15).

4.4.9 Parsimonious Cluster Algorithm

The goal of the Parsimonious Cluster algorithm is to partition all state contexts $i \in S^t$ of length t into an optimal set of disjoint equivalence classes. The general computational scheme of the Parsimonious Cluster algorithm has initially been proposed by Bourguignon and Robelin (2004) for higher-order MMs. Further studies and extensions of this computational scheme have been done by Gohr (2006). The basis behind the Parsimonious Cluster algorithm is a dynamic programming approach which efficiently computes the optimal set of disjoint equivalence classes. This is done by computing the tree score function $F(\tau_t(c))$ in (4.15) with the help of the extended tree ψ_t by successively adding the scores of optimal disjoint equivalence classes of state contexts. According to this, the Parsimonious Cluster algorithm is given by the following computational scheme.

- *Initialization*: For each leaf node v of the extended tree ψ_t in depth t consider the corresponding equivalence class ξ of v.
    1. Estimate the optimal transition probabilities $a_{\xi j}(c)^{(*)}$ for each next state $j \in S$, and store them in leaf node v.
        - Compute $a_{\xi j}(c)^{(*)}$ by (4.13) if state context length $1 \leq t < L$, otherwise use (4.14).
    2. Compute the score of the equivalence class $f_t(\vec{a}_\xi(c)^{(*)})$ in (4.10) based on the optimal transition probabilities $a_{\xi j}(c)^{(*)}$, and store this score in leaf node v.

- *Iteration*: Climb up one level towards the root. Consider each node v of the extended tree ψ_t in the current depth.
    1. Determine all child nodes of the current node v.
    2. Based on the labels of these child nodes, determine each combination of child nodes whose labels represent a partion ρ in the set of partitions Δ_S (4.2) for the set of hidden states S.
    3. Compute the score for each partition by adding the scores stored in the corresponding child nodes.

4. Parsimonious Higher-Order Hidden Markov Models

4. Determine the partition with the maximal score and store this score in the current node v.
5. Delete all sub-trees under v that have a root node which is not required for the partition with the maximal score.
6. Stop if the current node v is the root node of the extended tree ψ_t, otherwise continue with the next iteration step.

The Parsimonious Cluster algorithm iterates bottom-up from the leaf nodes to the root node of the extended tree ψ_t based on the initialization step and the iteration step.

The initialization step provides the basics for each equivalence class ξ of state contexts of length t contained in the set of all equivalence classes ξ_{ψ_t} defined in (4.16). For each equivalence class ξ the optimal transition parameters $a_{\xi j}(c)^{(*)}$ and the maximal score $f_t(\vec{a}_\xi(c)^{(*)})$ are computed. The equivalence class scoring function $f_t(\vec{a}_\xi(c)^{(*)})$ in (4.10) is the basic term of the tree score function $F(\tau_t(c))$ in (4.15). The tree score function itself is computed by the iteration steps towards the root of the extended tree.

The iteration step considers all nodes in the current depth. For each node v of these nodes the corresponding child nodes are evaluated. That is, the algorithm determines the child nodes of node v that define an optimal partition of the set of hidden states S with maximal score. Each combination of child nodes which define a partition $\rho \in \Delta_S$ based on their labels is considered. The score of each partition is computed by adding up the scores stored in the corresponding child nodes. The optimal partition of child nodes with maximal score remains under the parent node v, and the maximal score is stored in the parent node v. All sub-trees with root nodes not contained in the optimal partition are removed from the extended tree ψ_t. After the deletion of these sub-trees, the parent node v is the root node of an optimal sub-tree with maximal score. Each child node of v has child nodes whose set of labels defines an optimal partition with the highest score, because due to the previous iteration step each child node of v is itself the root of an optimal sub-tree with maximal score. That is, the optimal sub-tree with root node v is always constructed based on the optimal sub-trees of its child nodes. The fact that only the child nodes of the corresponding optimal partition are left under v ensures that each state $i \in S$ is represented by exactly one label of these child nodes. According to this, none of the state contexts of length t contained in ψ_t is lost by deleting all sub-trees under v with root nodes identical to child nodes that do not belong to the optimal partition.

The last iteration step considers only the root node of the tree ψ_t. All the $2^N - 1$ child nodes of the root node represent optimal sub-trees that have been successively

4. Parsimonious Higher-Order Hidden Markov Models

computed in the previous iteration steps. After removing all sub-trees with root nodes identical to child nodes that are not part of the optimal partition, the remaining extended tree ψ_t has been transformed into the optimal tree $\tau_t(c)^{(h+1)}$. The corresponding optimal transition probabilities $a_{\xi j}(c)^{(h+1)}$ are stored in the leaf nodes of the transformed extended tree ψ_t. Each leaf node of the transformed extended tree represents an equivalence class of state contexts of length t. All equivalence classes are disjoint, and all state contexts of length t are represented by ψ_t. The score contained in the root node of ψ_t is the maximal score that can be obtained by a set of disjoint equivalence classes under all different tree structures. This follows from the successive addition of optimal scores of disjoint equivalence classes during all iteration steps. This has been proven in a general form by Gohr (2006). That is, the obtained tree $\tau_t(c)^{(h+1)}$ maximizes the tree score function in (4.15) based on its optimal set of disjoint equivalence classes and the corresponding optimal transition probabilities.

The global objective function $F_t(A, \mathcal{T})$ in (4.9) is maximized for each state context length $1 \leq t \leq L$ by applying the Parsimonious Cluster algorithm to compute each optimal tree $\tau_t(c)^{(h+1)}$ and its corresponding optimal transition probabilities $a_{\xi j}(c)^{(h+1)}$. The resulting optimal trees and their corresponding optimal transition probabilities are assigned to the next inhomogeneous *PHHMM*(L, C) $\lambda(h+1)$, which is computed in the iteration step of the Bayesian Baum-Welch algorithm.

4.4.10 Computational Complexity of the Parsimonious Cluster Algorithm

The computational complexity of the Parsimonious Cluster algorithm is analyzed for the extended tree ψ_L of depth L. Based on this extended tree, the computation of the optimal tree $\tau_L(c)^{(h+1)}$ and its corresponding optimal transition probabilities $a_{\xi j}(c)^{(h+1)}$ for the next inhomogeneous *PHHMM*(L, C) $\lambda(h+1)$ has the highest computational complexity. This is because state contexts of the fixed maximal length L have to be considered. For the following analysis of the computational complexity, the *PHHMM*(L, C) is assumed to have N states and the considered part of an emission sequence comprises T emissions.

In the initialization step, each leaf node of the extended tree ψ_L is evaluated by computing the optimal transition probabilities $a_{\xi j}(c)^{(*)}$ and the optimal score $f_L(\vec{a}_\xi(c)^{(*)})$ for the equivalence class ξ of each leaf node. Since each non-leaf node of the extended tree ψ_L has exactly $2^N - 1$ child nodes, the initialization step has to operate on $(2^N - 1)^L$

4. Parsimonious Higher-Order Hidden Markov Models

leaf nodes. For each leaf node, the computation of the transition probability $a_{\xi j}(c)^{(*)}$ in (4.14) for the equivalence class ξ of a leaf node involves at most N^L different Epsilon-Variables for each time step t of the T time steps. For each of the $(2^N-1)^L$ equivalence classes represented by the leaf nodes of the extended tree, N different transition probabilities and the corresponding optimal score have to be computed. This leads to a run-time of $O\left((2^N-1)^L \cdot N^{L+1} \cdot T\right)$ for the initialization step.

The iteration step is working on each non-leaf node of the extended tree ψ_L. The total number of non-leaf nodes of this tree is $((2^N-1)^L - 1)/((2^N-1)-1)$. This follows from the geometric series that develops by the common ratio of $2^N - 1$ child nodes per non-leaf node. For each of these non-leaf nodes all different partitions of the set of hidden states S must be considered to compute the optimal partition and their corresponding optimal score. The number of different partitions is given by the Bell number $B[N]$ defined in (4.1). The computation of the score of each partition requires at most a sum over N scores stored in the child nodes of a non-leaf node. In addition to this, at most $2^N - 2$ sub-trees of child nodes that are not part of the optimal partition must be removed from each non-leaf node. This leads to a run-time of $O\left(((2^N-1)^L - 1)/((2^N-1)-1) \cdot (B[N] \cdot N + 2^N - 2)\right)$ that is mainly influenced by $B[N]$ which grows faster than 2^N for $N > 4$.

In summary, the upper bound of the run-time of the Parsimonious Cluster algorithm is given by the sum of the run-times of the initialization step and the iteration step.

5 Hidden Markov Models with Scaled Transition Matrices

The basis of the Hidden Markov Model with C Scaled transition matrices ($SHMM(C)$) is the standard homogeneous first-order *HMM* reviewed by Rabiner (1989). The $SHMM(C)$ extends this model by integrating additional information into the state-transition process. Based on this, the homogeneous first-order *MM* that underlies the *HMM* is substituted by a specific inhomogeneous first-order *MM* that represents the state-transition process of the $SHMM(C)$. The scaled transition matrices of the $SHMM(C)$ are specifically coupled to a basic transition matrix to realize increasing self-transition probabilities for the states of the $SHMM(C)$. This allows to model that two successive emissions can have a basic probability or an increased probability to be generated from the same state of the $SHMM(C)$ in dependency of integrated additional information for selecting one of the scaled transition matrices. The concept underlying the $SHMM(C)$ has initially been used by Seifert (2006) to analyze gene expression profiles of tumors in the context of distances between directly adjacent genes on a chromosome. In Seifert (2006) only the Baum-Welch algorithm that does not allow to integrate biological prior knowledge into the training has been considered. Here, the main focus of this chapter is on the extension of the $SHMM(C)$ to enable the integration of biological prior knowledge into the training by the usage of the Bayesian Baum-Welch algorithm.

Goals of this Chapter

1. The concept of scaled transition matrices underlying the $SHMM(C)$ is introduced.

2. The definition of the $SHMM(C)$ is given.

3. The Bayesian Baum-Welch algorithm for solving the *Maximum A Posteriori Problem* is adapted to the specific requirements of the $SHMM(C)$.

5. Hidden Markov Models with Scaled Transition Matrices

5.1 Scaling of Transition Matrices

The basis of transitions between the states of the standard homogeneous first-order HMM is the stochastic transition matrix $A := (a_{ij})$. Each element a_{ij} defines the probability for a transition from a current state $i \in S$ to a next state $j \in S$. In addition to this, the self-transition probability a_{ii} of state $i \in S$ characterizes the expected number of successive time steps which the HMM stays in state i before it goes to an another state $j \in S \setminus \{i\}$. This expected number of successive time steps is also referred to as the state duration

$$d_i := \frac{1}{1 - a_{ii}} \tag{5.1}$$

of state i, which is defined by the expectation value of the geometric distribution for staying in state i. Now, the goal is to derive a transition matrix $A_c := (a_{ij}(c))$ for each transition class $c \in C$ by increasing the state duration d_i in (5.1) for each state $i \in S$ by a pre-defined scaling factor $f_c \geq 1$. In the first step, the state duration d_i of state $i \in S$ in transition matrix A is multiplied with the scaling factor f_c leading to the state duration $d_i^{(c)} := f_c \cdot d_i$ of state i in transition matrix A_c. In analogy to (5.1), the state duration of state i in A_c can also be expressed by $d_i^{(c)} = 1/(1 - a_{ii}(c))$ in consideration of the self-transition probability $a_{ii}(c)$ in A_c. Next, both expressions for $d_i^{(c)}$ are used to compute $a_{ii}(c)$ resulting in

$$a_{ii}(c) = \frac{a_{ii} - 1 + f_c}{f_c} \tag{5.2}$$

which is depending on the self-transition probability a_{ii} of transition matrix A and the pre-defined scaling factor f_c. In the second step, the self-transition probability $a_{ii}(c)$ of state $i \in S$ and all non-self-transition probabilities $a_{ij}(c)$ to all states $j \in S \setminus \{i\}$ must fulfill the constraint $\sum_{j \in S} a_{ij}(c) = 1$. For that reason, all non-self-transition probabilities a_{ij} in transition matrix A for state i and for all next states $j \in S \setminus \{i\}$ are multiplied by a common factor $m_i^{(c)}$ to obtain each corresponding non-self-transition probability $a_{ij}(c) := m_i^{(c)} \cdot a_{ij}$ in transition matrix A_c. Thus, the constraint $\sum_{j \in S} a_{ij}(c) = 1$ is rewritten to $a_{ii}(c) + m_i^{(c)} \cdot \sum_{j \in S \setminus \{i\}} a_{ij} = 1$, which is finally transformed to $a_{ii}(c) + m_i^{(c)} \cdot (1 - a_{ii}) = 1$ using the constraint $\sum_{j \in S \setminus \{i\}} a_{ij} = 1 - a_{ii}$. Based on this, $m_i^{(c)}$ is computed by substituting $a_{ii}(c)$ with its expression in (5.2) resulting in the common factor $m_i^{(c)} = 1/f_c$.

5. Hidden Markov Models with Scaled Transition Matrices

In consideration of this common factor, each non-self-transition probability

$$a_{ij}(c) = \frac{a_{ij}}{f_c} \qquad (5.3)$$

for state $i \in S$ and state $j \in S \setminus \{i\}$ in transition matrix A_c is depending on its corresponding entry a_{ij} in matrix A and the pre-defined scaling factor f_c. Taking this together, the transition matrix A_c with increased state durations in comparison to the basic matrix A is defined by

$$A_c := (a_{ij}(c)) := \begin{cases} \frac{a_{ii}-1+f_c}{f_c} & , i = j \\ \frac{a_{ij}}{f_c} & , i \neq j \end{cases} \qquad (5.4)$$

on the basis of the pre-defined scaling factor $f_c \geq 1$ and the given basic stochastic transition matrix A.

5.2 Hidden Markov Model with Scaled Transition Matrices

The *SHMM(C)* with C scaled transition matrices is defined by $\lambda = (\vec{\pi}, \vec{f}, A, B)$ based on the following parameters.

1. The initial state distribution $\vec{\pi} := (\pi_{S_1}, \ldots, \pi_{S_N})$ defines for each state $i \in S$ the probability $\pi_i := P[Q_1 = i]$ of starting in this state at time step $t = 1$. Two stochastic constraints must be fulfilled by $\vec{\pi}$.
   a) $\forall i \in S : \pi_i \in [0, 1]$
   b) $\sum_{i \in S} \pi_i = 1$

2. The vector of scaling factors $\vec{f} := (f_1, \ldots, f_C)$ with $f_1 := 1$ and $f_1 < f_2 < \ldots < f_C$ to scale the state durations in the basic transition matrix A.

3. The basic transition matrix $A := (a_{ij})$ defines for each transition from state $i \in S$ to state $j \in S$ the corresponding transition probability a_{ij}. Each row $i \in S$ of A must fulfill the following two stochastic constraints.
   a) $\forall j \in S : a_{ij} \in [0, 1]$
   b) $\sum_{j \in S} a_{ij} = 1$

5. Hidden Markov Models with Scaled Transition Matrices

Based on this matrix A and the vector of scaling factors $\vec{f}$, the definition in (5.4) is used to obtain the C scaled transition matrices $A_1, \ldots, A_C$. It follows that A_1 is identical to the basic transition matrix A because of $f_1 := 1$. Generally, the transition matrix $A_c := (a_{ij}(c))$ defines the transition probability $a_{ij}(c) := P[Q_{t+1} = j \mid Q_t = i, c]$ for a transition from the current state $i \in S$ to the next state $j \in S$ in consideration of the given transition class c at a specific time step t.

4. The matrix $B := (\mu_i, \sigma_i)$ defines the state-specific mean $\mu_i \in \mathbb{R}$ and the state-specific standard deviation $\sigma_i \in \mathbb{R}^+$ for the Gaussian emission density of each state $i \in S$. The time-independent probability density $b_i(o) := P[O_t = o \mid Q_t = i]$ for emitting an emission $o \in \mathbb{R}$ by the Gaussian emission density of state i is defined in (3.2).

With respect to the notation scheme in Tab. 3.1, the $SHMM(C)$ simplifies to the homogeneous HMM for $C = 1$ transition class. Due to the specific mapping in (5.4) for obtaining the scaled transition matrices, the $SHMM(C)$ can be considered as a special case of the inhomogeneous $HMM(C)$. The $HMM(C)$ itself represents a special case of the inhomogeneous $HHMM(L,C)$ of order $L = 1$. Thus, the Forward algorithm and the Backward algorithm, which both provide the basics for other computations, and the Viterbi algorithm can be used for the $SHMM(C)$ without any modifications.

5.3 Solving the Maximum A Posteriori Problem

The basis to solve the *Maximum A Posteriori Problem* for the $SHMM(C)$ has already been described in detail for the inhomogeneous $HHMM(L,C)$ in Sec. 3.7.1. The Bayesian Baum-Welch algorithm developed there to solve this problem only requires the adaptation of the transition parameter estimation due to the specific modeling of self-transition probabilities by the $SHMM(C)$. For that reason, Baum's auxiliary function for transition parameters in (3.22) and the corresponding transition prior in (3.31) have to be modified for the $SHMM(C)$. The estimation of the start probabilities and of the emission parameters does not need to be adapted. These estimations are done with respect to the current $SHMM(C)$ $\lambda(h)$ of iteration step h as specified in the Bayesian Baum-Welch algorithm in Sec. 3.7.3. Subsequently, the transition prior and Baum's auxiliary function for transition parameters are modified, and the non-self-transition probabilities and the self-transition probabilities of the next $SHMM(C)$ $\lambda(h+1)$ are determined.

5. Hidden Markov Models with Scaled Transition Matrices

5.3.1 Transition Prior

The transition prior $D_2^t(A | \Theta_2)$ defined in (3.31) for the transition parameters of the inhomogeneous *HHMM(L, C)* is adapted to a product of transformed Dirichlet distributions

$$D_2(A | \Theta_2) := \prod_{i \in S} Z(\Theta_2^i) \prod_{j \in S} \exp\left(\Lambda_{a_{ij}} \cdot \vartheta_{ij}\right) \quad (5.5)$$

to define the prior for the transition matrix A of the *SHMM(C)* with respect to the relation $a_{ij} := \exp(\Lambda_{a_{ij}})$. Here, the matrix $\Theta_2 := (\Theta_2^{S_1}, \ldots, \Theta_2^{S_N})$ defines based on the vector $\Theta_2^i := (\vartheta_{iS_1}, \ldots, \vartheta_{iS_N})$ with $\vartheta_{ij} \in \mathbb{R}^+$ the prior knowledge for each transition from state $i \in S$ to state $j \in S$. Again, the corresponding normalization constant for each Dirichlet distribution is defined by $Z(\Theta_2^i) := \Gamma(\sum_{j \in S} \vartheta_{ij}) / \prod_{j \in S} \Gamma(\vartheta_{ij})$ in consideration of the Gamma function $\Gamma(x) = \int_0^\infty u^{x-1} \cdot \exp(-u) \, du$ for all $x \in \mathbb{R}^+$.

5.3.2 Baum's auxiliary function for Transition Parameters

Baum's auxiliary function $Q_2^L(A | \lambda(h))$ defined in (3.22) for the transition parameters of the inhomogeneous *HHMM(L, C)* provides the basics to derive this function for the specific transition probabilities of the *SHMM(C)*. In analogy, Baum's auxiliary function to estimate the transition probabilities of the next *SHMM(C)* $\lambda(h+1)$ is initially given by

$$Q_2(A | \lambda(h)) := \sum_{c \in C} \sum_{i \in S} \sum_{j \in S} \log(a_{ij}(c)) \sum_{k=1}^{K} \sum_{\substack{t=1 \\ c_t(k) = c}}^{T_k - 1} \varepsilon_t^k(i, j)$$

in consideration of the Epsilon-Variables $\varepsilon_t^k(i, j)$ in (3.20) computed under the current *SHMM(C)* $\lambda(h)$. In order to account for the scaled self-transition probability $a_{ii}(c)$ in (5.2) and to account for the non-self-transition probability $a_{ij}(c)$ in (5.3), Baum's auxiliary function for transition parameters is splitted into two terms represented by

$$Q_2(A | \lambda(h)) = \sum_{c \in C} \sum_{i \in S} \log\left(\frac{\exp(\Lambda_{a_{ii}}) - 1 + f_c}{f_c}\right) \sum_{k=1}^{K} \sum_{\substack{t=1 \\ c_t(k) = c}}^{T_k - 1} \varepsilon_t^k(i, i)$$

$$+ \sum_{c \in C} \sum_{i \in S} \sum_{j \in S \setminus \{i\}} \log\left(\frac{\exp(\Lambda_{a_{ij}})}{f_c}\right) \sum_{k=1}^{K} \sum_{\substack{t=1 \\ c_t(k) = c}}^{T_k - 1} \varepsilon_t^k(i, j) \quad (5.6)$$

75

5. Hidden Markov Models with Scaled Transition Matrices

under additional consideration of the parameterization $a_{ii} := \exp(\Lambda_{a_{ii}})$ and $a_{ij} := \exp(\Lambda_{a_{ij}})$. The first term represents the self-transition probabilities and the second term considers the corresponding non-self-transition probabilities. Based on that, both types of transition probabilities of the next $SHMM(C)$ are determined subsequently.

5.3.3 Estimation of Transition Parameters

In analogy to the transition parameter estimation for the Bayesian Baum-Welch algorithm of the inhomogeneous $HHMM(L,C)$, the standard Lagrange optimization technique (e.g. Bishop (2006)) is used to provide the basics for the transition parameter estimation of the $SHMM(C)$. Besides this, the application of a numerical optimization method like Newton's method is necessary to compute the transition probabilities of the $SHMM(C)$. The basis of the transition parameter estimation is given by the auxiliary function

$$H(A\,|\,\lambda(h)) := Q_2(A\,|\,\lambda(h)) + \log(D_2(A\,|\,\Theta_2)) - \sum_{i \in S} \delta_i \left(\left(\sum_{j \in S} \exp(\Lambda_{a_{ij}}) \right) - 1 \right) \quad (5.7)$$

in consideration of Baum's auxiliary function for transition parameters $Q_2(A\,|\,\lambda(h))$ in (5.6), the transition prior $D_2(A\,|\,\Theta_2)$ in (5.5), and the constraint $\sum_{j \in S} \exp(\Lambda_{a_{ij}}) = 1$ in combination with the Lagrange multiplier δ_i. That is, $H(A\,|\,\lambda(h))$ is specified in analogy to the auxiliary function used for the transition parameter estimation of the inhomogeneous $HHMM(L,C)$. Subsequently, the focus is on the computation of the non-self-transition probabilities of the next $SHMM(C)$ $\lambda(h+1)$. Based on this, the basics for the computation of the self-transition probabilities by Newton's method are provided.

Non-Self-Transition Probabilities

To compute the basic non-self-transition probability $a_{ij}^{(h+1)}$ for a transition from a fixed current state $i \in S$ to a next state $j \in S \setminus \{i\}$ under the next $SHMM(C)$ $\lambda(h+1)$ the auxiliary function $H(A\,|\,\lambda(h))$ in (5.7) has to be maximized. That is, $H(A\,|\,\lambda(h))$ is first differentiated with respect to $\Lambda_{a_{ij}}$ and with respect to the Lagrange multiplier δ_i. Next, the resulting derivatives are set equal to zero to obtain the non-self-transition probability $a_{ij}^{(h+1)}$ based on the relation $a_{ij} = \exp(\Lambda_{a_{ij}})$. Here, this basic non-self-transition

5. Hidden Markov Models with Scaled Transition Matrices

probability for a transition from state i to state $j \neq i$ is given by

$$a_{ij}^{(h+1)} = \frac{1}{\delta_i}\left(\left(\sum_{c \in \mathcal{C}}\sum_{k=1}^{K}\sum_{\substack{t=1 \\ c_t(k)=c}}^{T_k-1}\varepsilon_t^k(i,j)\right) + \vartheta_{ij}\right) \quad (5.8)$$

in dependency of the corresponding Lagrange multiplier

$$\delta_i = \frac{1}{1 - a_{ii}^{(h+1)}}\left(\sum_{j \in S\setminus\{i\}}\left(\sum_{c \in \mathcal{C}}\sum_{k=1}^{K}\sum_{\substack{t=1 \\ c_t(k)=c}}^{T_k-1}\varepsilon_t^k(i,j)\right) + \vartheta_{ij}\right) \quad (5.9)$$

to ensure the constraint $\sum_{j \in S} a_{ij} = a_{ii} + \sum_{j \in S\setminus\{i\}} a_{ij} = 1$ in (5.7). In both cases, the Epsilon-Variables $\varepsilon_t^k(i,j)$ defined in (3.20) are computed under the current $SHMM(C)$ $\lambda(h)$, and ϑ_{ij} is specified by the transition prior in (5.5). In addition to this, the non-self-transition probability $a_{ij}^{(h+1)}$ is depending on the corresponding self-transition probability $a_{ii}^{(h+1)}$ due to the expression of the Lagrange multiplier in (5.9). Using the formula (5.3) based on the expressions for $a_{ij}^{(h+1)}$ in (5.8) and δ_i in (5.9), the corresponding scaled non-self-transition probability

$$a_{ij}(c)^{(h+1)} = \frac{(1 - a_{ii}^{(h+1)}) \cdot \left(\left(\sum_{c \in \mathcal{C}}\sum_{k=1}^{K}\sum_{\substack{t=1 \\ c_t(k)=c}}^{T_k-1}\varepsilon_t^k(i,j)\right) + \vartheta_{ij}\right)}{f_c \cdot \left(\sum_{j \in S\setminus\{i\}}\left(\sum_{c \in \mathcal{C}}\sum_{k=1}^{K}\sum_{\substack{t=1 \\ c_t(k)=c}}^{T_k-1}\varepsilon_t^k(i,j)\right) + \vartheta_{ij}\right)} \quad (5.10)$$

for transition class $c \in \mathcal{C}$ is obtained. Subsequent to this, the basics to compute the corresponding self-transition probability $a_{ii}^{(h+1)}$ are given.

Self-Transition Probabilities

The first step to determine the basic self-transition probability $a_{ii}^{(h+1)}$ of state $i \in S$ under the next $SHMM(C)$ $\lambda(h+1)$ is to differentiate the auxiliary function $H(A \mid \lambda(h))$ in (5.7) with respect to $\Lambda_{a_{ii}}$. Next, the resulting derivation is set to zero to derive the optimal

5. Hidden Markov Models with Scaled Transition Matrices

$a_{ii}^{(h+1)}$ based on the relation $a_{ii} = \exp(\Lambda_{a_{ii}})$. This leads to the following equation

$$\left(\sum_{c \in C} \frac{1}{a_{ii}^{(h+1)} - 1 + f_c} \sum_{k=1}^{K} \sum_{\substack{t=1 \\ c_t(k) = c}}^{T_k - 1} \varepsilon_t^k(i,i) \right) + \frac{\vartheta_{ii}}{a_{ii}^{(h+1)}} = \delta_i$$

for which the right-hand side is defined by the specific expression of the Lagrange multiplier δ_i of state i in (5.9). This equation is multiplied by $1 - a_{ii}^{(h+1)}$ and then the resulting term $(1 - a_{ii}^{(h+1)})/(a_{ii}^{(h+1)} - 1 + f_c)$ on the left-hand side is substituted by its equivalent expression $f_c/(a_{ii}^{(h+1)} - 1 + f_c) - 1$. Finally, the Lagrange multiplier δ_i is substituted by its corresponding expression in (5.9) and the missing term for a sum over $j \in S$ instead of a sum over $j \in S \setminus \{i\}$ is added from the left-hand side to the right-hand side. This leads to the final equation $L(a_{ii}^{(h+1)}) = R_i$ in which the left-hand side is given by the function

$$L(a_{ii}^{(h+1)}) := \left(\sum_{c \in C} \frac{f_c}{a_{ii}^{(h+1)} - 1 + f_c} \sum_{k=1}^{K} \sum_{\substack{t=1 \\ c_t(k) = c}}^{T_k - 1} \varepsilon_t^k(i,i) \right) + \frac{\vartheta_{ii}}{a_{ii}^{(h+1)}}$$

depending on the value of $a_{ii}^{(h+1)}$ for the next *SHMM(C)* $\lambda(h+1)$. The corresponding right-hand side

$$R_i := \sum_{j \in S} \left(\left(\sum_{c \in C} \sum_{k=1}^{K} \sum_{\substack{t=1 \\ c_t(k) = c}}^{T_k - 1} \varepsilon_t^k(i,j) \right) + \vartheta_{ij} \right)$$

is a constant. For both terms, the Epsilon-Variables $\varepsilon_t^k(i,j)$ defined in (3.20) are computed under the current *SHMM(C)* $\lambda(h)$ and each $\vartheta_{ij} \in \mathbb{R}^+$ is specified for the transition prior in (5.5).

In the general case of C scaled transition matrices, the final equation $L(a_{ii}^{(h+1)}) = R_i$ cannot be solved analytically. However, the following characteristics of $L(a_{ii}^{(h+1)})$ and R_i allow to solve $L(a_{ii}^{(h+1)}) = R_i$ numerically.

1. $L(a_{ii}^{(h+1)})$ is strictly monotonic decreasing for increasing values of $a_{ii}^{(h+1)}$ in the interval $(0,1)$ based on the fact that $L(a_{ii}^{(h+1)})$ is a sum of hyperbolas.

2. The lower limit of $L(a_{ii}^{(h+1)})$ is $\sum_{c \in C} \sum_{k=1}^{K} \sum_{\substack{t=1 \\ c_t(k) = c}}^{T_k - 1} \varepsilon_t^k(i,i) + \vartheta_{ii}$ for $a_{ii}^{(h+1)} = 1$.

5. Hidden Markov Models with Scaled Transition Matrices

3. $L(a_{ii}^{(h+1)})$ grows against infinity if $a_{ii}^{(h+1)}$ is decreased towards zero.

4. R_i is always greater than the lower limit of $L(a_{ii}^{(h+1)})$ due to the fact that the lower limit of $L(a_{ii}^{(h+1)})$ is already included in R_i for $j = i$ in combination with the fact that all Epsilon-Variables $\varepsilon_t^k(i,j)$ are greater than zero.

That means, it is always possible to find exactly one $a_{ii}^{(h+1)} \in (0,1)$ that solves the equation $L(a_{ii}^{(h+1)}) = R_i$. A good strategy to compute this $a_{ii}^{(h+1)}$ is to apply Newton's method. That is, based on an initial choice of $a_{ii}^{(h+1)}[0] \in (0,1)$ improved values are computed iteratively by

$$a_{ii}^{(h+1)}[u+1] := a_{ii}^{(h+1)}[u] - \frac{L(a_{ii}^{(h+1)}[u]) - R_i}{L'(a_{ii}^{(h+1)}[u])}$$

in consideration of the function $L'(a_{ii}^{(h+1)})$, which is the derivative of $L(a_{ii}^{(h+1)}) - R_i$ with respect to $a_{ii}^{(h+1)}$. This derivative is given by

$$L'(a_{ii}^{(h+1)}[u]) = -\left(\sum_{c \in \mathcal{C}} \frac{f_c}{\left(a_{ii}^{(h+1)}[u] - 1 + f_c\right)^2} \sum_{k=1}^{K} \sum_{\substack{t=1 \\ c_t(k) = c}}^{T_k - 1} \varepsilon_t^k(i,i) \right) - \frac{\vartheta_{ii}}{\left(a_{ii}^{(h+1)}[u]\right)^2}$$

with respect to the current value of $a_{ii}^{(h+1)}[u]$. The final $a_{ii}^{(h+1)}$ obtained through the application of Newton's method is the desired self-transition probability of state $i \in S$ of the next $SHMM(C)$ $\lambda(h+1)$. This $a_{ii}^{(h+1)}$ is used to compute each non-self-transition probability $a_{ij}(c)^{(h+1)}$ in (5.10). Additionally, $a_{ii}^{(h+1)}$ is used to compute the corresponding scaled self-transition probability $a_{ii}(c)^{(h+1)}$ in (5.2) for each transition class $c \in \mathcal{C}$. In summary, all these computations and the resulting estimation of the self-transition probability $a_{ii}(c)^{(h+1)}$ and of the non-self-transition probability $a_{ij}(c)^{(h+1)}$ are done in the iteration step of the Bayesian Baum-Welch algorithm in Sec. 3.7.3 with respect to the current $SHMM(C)$ $\lambda(h)$.

6 Analysis of Breast Cancer Gene Expression Data

Chromosomal mutations like amplifications and deletions of DNA segments are one of the key genetic mechanisms that lead to changes of gene expression levels in tumors. Different studies have shown that between 40% and 60% of the genes in highly amplified regions tend to be over-expressed (Hyman et al. (2002); Pollack et al. (2002); Heidenblad et al. (2005)). In addition to this, also long-range epigenetic changes of DNA methylations or histone modifications are known to bias expression levels in chromosomal regions (Frigola et al. (2006); Stransky et al. (2006)). Due to such mutations gene expression levels of adjacent genes in close chromosomal proximity tend to be higher correlated than those of adjacent genes in greater distance.

In recent years different approaches have been proposed to analyze gene expression data in the context of chromosomal locations. The Human Transcriptome Map by Caron et al. (2001) was the first large-scale approach to study genome-wide human gene expression profiles in their chromosomal context. The mapping of gene expression data to corresponding chromosomal locations revealed a higher order organization of the human genome in which highly expressed genes tend to be localized in clusters. In addition to this, methods like CGMA (Comparative Genomic Microarray Analysis) by Crawley and Furge (2002), MACAT (Microarray Chromosome Analysis Tool) by Toedling et al. (2004), or LAP (Locally Adaptive statistical Procedure) by Callegaro et al. (2006) have been developed to improve the analysis of gene expression data in the context of chromosomal locations. However, a common characteristic of all these methods is the comparison of two defined samples, e.g. tumor tissue against healthy tissue, based on specific test statistics coupled with permutation tests to identify differentially expressed chromosomal regions. Thus, these methods cannot be applied to data sets for which two defined samples do not exist. This includes the breast cancer data set by Pollack et al. (2002) which is based on two-color microarrays to measure the relative difference of gene expression levels in two samples.

6. Analysis of Breast Cancer Gene Expression Data

Here, the goal is to develop methods for analyzing data sets like that of Pollack et al. (2002) by *HMM*-based approaches that integrate chromosomal locations or chromosomal distances of genes. Motivated through the histogram of log-ratios and the quantile-quantile plot in Fig. 6.1, a three-state *HMM* is introduced for the analysis of breast cancer gene expression data in the context of chromosomal locations of genes. This model is extended to a *SHMM*(2) with two scaled transition matrices that integrates chromosomal distances of directly adjacent genes on a chromosome. This extension is motivated by the trend shown in Fig. 6.2 for the breast cancer gene expression data set by Pollack et al. (2002). That is, adjacent genes in close chromosomal proximity tend to have higher correlated expression levels than adjacent genes in greater distance. Based on these two models, the effect of integrating biological prior knowledge into the training of the *HMM* and the *SHMM*(2) is investigated by comparing the Bayesian Baum-Welch algorithm against the standard Baum-Welch algorithm. Beyond that, the *HMM* and the *SHMM*(2) are compared to related methods that have been developed for the analysis of array comparative genomic hybridization (Array-CGH) experiments. The technique that underlies these experiments is reviewed in a general form by Pollack et al. (1999) and Pinkel and Albertson (2005). All methods comprising the *HMM*, the *SHMM*(2) and those of the Array-CGH field are initially applied to the breast cancer gene expression data set by Pollack et al. (2002). The best performing methods are further validated by predicting the direct effects of amplifications and deletions on the gene expression levels in breast cancer. Genes frequently predicted as under-expressed or over-expressed are checked for their occurrence in two independent public databases, the Genetic Association Database (Becker et al. (2004)) and the Breast Cancer Database (Telikicherla et al. (2008)). These databases collect genes that are known to play a role in different types of breast cancer. Those genes that are not contained in both databases are further investigated for their role in breast cancer by additional literature searches. Besides this, the influence of integrating chromosomal locations and chromosomal distances of genes on the prediction of under-expressed and over-expressed genes is investigated in more detail by the comparison to a mixture model that does not integrate these additional information.

Goals of this Chapter

1. The *HMM* and the *SHMM*(2) used for the analysis of breast cancer gene expression data are developed.

2. Related methods from the field of Array-CGH data analysis that are tested on the breast cancer gene expression data set are summarized.

3. The effect of using the Baum-Welch algorithm or the Bayesian Baum-Welch algorithm for the training of the *HMM* and the *SHMM*(2) is investigated for the prediction of differentially expressed genes in breast cancer.

4. All methods including the *HMM*, the *SHMM*(2), and the related methods from the field of Array-CGH data analysis are analyzed to find out which methods are appropriate for the analysis of breast cancer gene expression data.

5. The effect of modeling chromosomal distances of genes on the self-transition probabilities of the *SHMM*(2) is analyzed.

6. The *HMM*, the *SHMM*(2), and GLAD are validated by predicting the direct effects of amplifications and deletions on the gene expression levels in breast cancer.

7. The influence of modeling chromosomal locations and distances of genes on the prediction of under-expressed and over-expressed genes in breast cancer is investigated.

8. Genes frequently predicted as over-expressed or under-expressed are further investigated using independent public databases. Genes not included in these databases are further investigated by additional literature searches.

6.1 Breast Cancer Gene Expression Data Set

The breast cancer gene expression data set created by Pollack et al. (2002) is used to predict genes that are differentially expressed in breast cancer in comparison to a pool of reference cell lines (Perou et al. (2000)). This data set contains gene expression levels for 4 breast cancer cell lines and 37 tumors across 6,095 genes of the 23 human chromosomes. The T_k gene expression levels of a chromosome k that have been measured for a cell line or a tumor are represented by an emission sequence $\vec{o}(k) = (o_1(k), \ldots, o_{T_k}(k))$. Here, $o_t(k)$ is defined to be the log-ratio of the gene expression level of gene t on chromosome k in tumor in relation to the gene expression level of this gene in the pool of reference cell lines. All log-ratios in an emission sequence are sorted from the p-arm to the q-arm of the chromosome based on the chromosomal

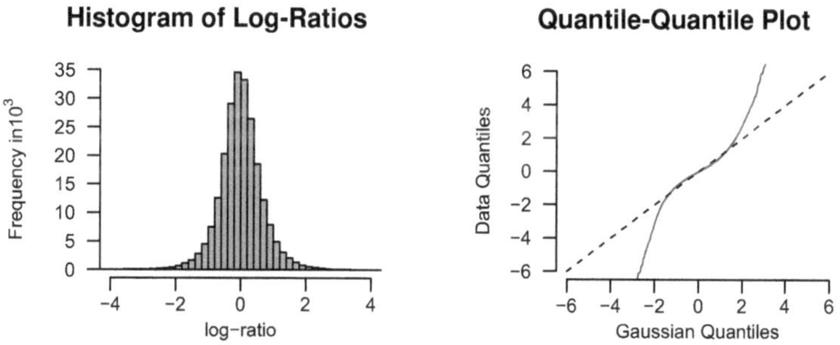

6. Analysis of Breast Cancer Gene Expression Data

Figure 6.1: Overview of the breast cancer gene expression data set by Pollack et al. (2002) showing the histogram of log-ratios and the quantile-quantile plot. The histogram of log-ratios represents the 6,095 genes that have been measured for each of the 4 breast cancer cell lines and for each of the 37 tumors. The log-ratio of a gene represents the ratio between the expression level of this gene in tumor compared to the expression level of this gene in reference cell lines. Genes with unchanged expression levels in tumor have log-ratios about zero, and differentially expressed genes are expected to have log-ratios much less (under-expressed) or much greater (over-expressed) than zero. The quantile-quantile plot characterizes the quantiles of the data set with respect to the quantiles obtained for the Gaussian density with mean 0.01 and standard deviation 0.7 estimated from the data set. Log-ratios with values much different from zero are represented clearly more frequently in the data set in comparison to the estimated Gaussian density.

locations of the corresponding genes provided by Pollack et al. (2002). An overview to the breast cancer gene expression data set is shown in Fig. 6.1.

6.2 Methods for Breast Cancer Gene Expression Data Analysis

6.2.1 Hidden Markov Model approach

Model: An *HMM* with three states $S := \{-, =, +\}$ and state-specific Gaussian emission densities is used to identify differentially expressed genes in breast cancer. The *HMM* of order $L = 1$ with one transition class $C = 1$ is a special case of the $HHMM(L, C)$ defined in Sec. 3.2. The state '=' represents genes with unchanged expression levels between tumor and the pool of reference cell lines characterized in Fig. 6.1 by log-ratios with values about zero. Differentially expressed genes are

6. Analysis of Breast Cancer Gene Expression Data

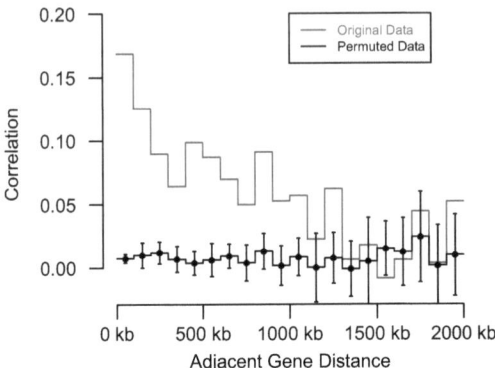

Figure 6.2: Connection between distances of directly adjacent genes on chromosomes and correlation of their log-ratios for the breast cancer gene expression data set by Pollack et al. (2002). Adjacent genes on each chromosome have been grouped initially into distance classes between 100 kb and to 2000 kb in steps of 100 kb. Pearson correlations for the log-ratios of all pairs of adjacent genes in one distance class have been computed for the original breast cancer gene expression data and for 100 random permutations of log-ratios per chromosome of the original data set.

modeled by the two states '−' and '+'. Here, state '−' represents under-expressed genes in tumor that are characterized by log-ratios less than zero, and state '+' models over-expressed genes in tumor with log-ratios greater than zero. The *HMM* models dependencies between measurements of directly adjacent genes on a chromosome based on its underlying first-order Markov chain. The chromosomal distance between directly adjacent genes cannot be integrated into the *HMM*. Thus, the trend shown in Fig. 6.2 indicating that the positive correlation of log-ratios of two adjacent genes tends to decrease with increasing chromosomal distance of genes cannot be modeled by the *HMM*. Yet, the *HMM* is an important reference for the comparison to the *SHMM*(2) that integrates chromosomal distances of genes.

Prior: A good biological characterization of each state can be achieved by including prior knowledge into the training of the model. The histogram of log-ratios and the quantile-quantile plot in Fig. 6.1 show that it can be expected that under-expressed genes in tumor have log-ratios less than zero, genes with unchanged expression levels between tumor and the reference cell lines have log-ratios about zero, and over-

6. Analysis of Breast Cancer Gene Expression Data

expressed genes in tumor have log-ratios greater than zero. For that reason, the means of the Gaussian distributions of the emission prior are set to $\eta_- = -2$, $\eta_= = 0$, and $\eta_+ = 2$ to distinguish differentially expressed genes from unchanged expressed genes in tumor. Additionally, it is expected that differentially expressed genes have log-ratios much less or much greater than zero. Based on that, less flexibility is allowed for the training of the emission parameters of the states '−' and '+' by using the scale parameters $\epsilon_- = \epsilon_+ = 15,000$, and more flexibility is given to the state '=' by using $\epsilon_= = 1,000$. To enhance the separation between putative under-expressed, unchanged expressed, and over-expressed genes in tumor, the values of $r_i = 10$ and $\alpha_i = 10^{-4}$ are used for each state $i \in S$ to obtain proper standard deviations. Finally, the same start prior parameter $\vartheta_i = 3$ is assigned to each state $i \in S$ for its initial state probability, and the transition prior parameter $\vartheta_{ij}(1) = 1$ is used for the corresponding transition probabilities.

Initialization: The initial *HMM* must be able to differentiate between differentially expressed genes and genes with unchanged expression levels in tumor. Again, the histogram of log-ratios and the quantile-quantile plot in Fig. 6.1 help to choose initial model parameters. First the initial state probabilities of the three states are chosen. The proportion of under-expressed and the proportion of over-expressed genes is much less than the proportion of unchanged expressed genes. Thus, the initial state distribution is set to $\vec{\pi} = (0.1, 0.8, 0.1)$ using $\pi_- = \pi_+ = 0.1$ and $\pi_= = 0.8$. Based on that, the initial transition matrix $A_1 = (a_{ij}(1))_{i,j \in S}$ is chosen to have the stationary distribution identical to $\vec{\pi}$ by setting all diagonal elements to $a_{ii}(1) = 1 - s/\pi_i$ and all non-diagonal elements to $a_{ij}(1) = s/(2\pi_i)$ using $s = 0.05$ to control the state durations. In addition to this, all three states are characterized by proper means and standard deviations to represent the log-ratios. Here, the means $\mu_- = -2$, $\mu_= = 0$, and $\mu_+ = 2$ and the corresponding standard deviations $\sigma_- = 0.3$, $\sigma_= = 0.5$, and $\sigma_+ = 0.3$ are used.

Training: The initial *HMM* is trained with all emission sequences using the Bayesian Baum-Welch algorithm developed in Sec. 3.7.1. In addition to this, the initial *HMM* is also trained using the standard Baum-Welch algorithm given in Sec. 3.5 to compare the models obtained by the two training algorithms. In both cases, the training is stopped if the increase of the optimization function is less than 10^{-9} for two successive steps.

Detection of differentially expressed genes: For the initial comparison of the *HMM* to other methods the Viterbi algorithm described in Sec. 3.4.2 is used to assign one of the three state labels '−', '=', or '+' to each log-ratio in an emission sequence. The detailed investigation of the prediction results of the *HMM* is done by computing a score

6. Analysis of Breast Cancer Gene Expression Data

that measures the potential of a gene to be under-expressed or over-expressed. That is, the score $1-\gamma_t^k('=')$ of gene t in emission sequence $\vec{o}(k)$ is computed with respect to the probability that this gene is modeled by state '=' as unchanged expressed given by the state-posterior $\gamma_t^k('=')$ in (3.6). Based on this score, genes can be ranked for comparisons to other methods. Each gene that is considered as differentially expressed is labeled as under-expressed if it has a negative log-ratio, and otherwise this gene is considered as over-expressed.

6.2.2 Hidden Markov Model with two scaled transition matrices

Model: A *SHMM*(2) with three states $S := \{-, =, +\}$ and state-specific Gaussian emission densities, as shown in Fig. 6.3, is used to predict differentially expressed genes in breast cancer gene expression data by integrating chromosomal distances between adjacent genes on chromosomes. Again, the state '=' represents genes with unchanged expression levels between tumor and the reference cell lines, under-expressed genes are modeled by state '−', and state '+' represents over-expressed genes in tumor. Now, the trend that the positive correlation of log-ratios of two adjacent genes decreases with increasing chromosomal distance of genes (Fig. 6.2) is integrated into the *SHMM*(2). Thus, two adjacent genes with chromosomal distance less or equal than a pre-defined distance threshold $b \in \mathbb{N}$ have a higher probability to be represented by the same state of the *SHMM*(2) than two adjacent genes with chromosomal distance greater than b. To model this, two scaled transition matrices A_1 and A_2 are used. These two transition matrices are computed like specified in Sec. 5.2 using the basic transition matrix A, the pre-defined scaling factor $f_1 := 1$, and the user-defined scaling factor $f_2 > f_1$ of the *SHMM*(2). Here, the basic transition matrix A is initially set like the matrix A_1 of the previously defined *HMM*. To define the corresponding transition matrix for each pair of two adjacent genes t and $t+1$ on a chromosome k, the transition class

$$c_t(k) := \begin{cases} 2, & \text{genes } t \text{ and } t+1 \text{ have distance } d_t \leq b \\ 1, & \text{otherwise} \end{cases}$$

is assigned to each pair of adjacent genes in dependence of the chromosomal distance between both genes and the globally defined distance threshold b. Based on that, the *SHMM*(2) transitions from the state of gene t to the state of gene $t+1$ by using the corresponding transition matrix $A_{c_t(k)}$. Adjacent gene pairs modeled by A_2

have a higher probability to be represented by the same state of the SHMM(2) than gene pairs modeled by A_1, because the self-transition probability of each state $i \in S$ in A_2 is greater than in A_1. In the following, gene pairs modeled by A_2 are referred to as near gene pairs and gene pairs modeled by A_1 are referred to as far gene pairs.

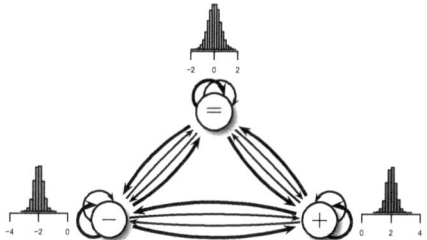

Figure 6.3: The basic three-state architecture of the SHMM(2) with two scaled transition matrices that is used for the analysis of breast cancer gene expression data. The states $S := \{-, =, +\}$ are represented by labeled circles and corresponding state-specific Gaussian emission densities. Chromosomal distances between adjacent genes on a chromosome are integrated into the SHMM(2) by modeling adjacent genes in close chromosomal proximity and adjacent genes in greater chromosomal distance. Thick arrows represent specific transitions for adjacent genes in close chromosomal proximity and thin arrows represent those of adjacent genes in greater chromosomal distance.

Prior: For the SHMM(2) the same prior settings are used as previously specified for the HMM.

Initialization: The basic initialization of the SHMM(2) is done like previously described for the HMM. In addition to this, the global distance threshold b and the scaling factor f_2 have to be specified. The relation between the distances of adjacent genes on a chromosome and the correlations of their log-ratios shown in Fig. 6.2 can help to find appropriate settings. Here, the log-ratios of adjacent genes in distances greater than 1000 kb showed generally only weak positive correlations that are comparable to those obtained under permuted data. For that reason, the maximal distance threshold is set to 1000 kb. The scaling factor f_2 allows to adjust the probability that two directly adjacent genes are represented by the same state of the SHMM(2). To high values of f_2 could lead to undesired predictions like pairs of over-expressed genes in which one gene has a log-ratio slightly less than zero. For these reasons, each global distance threshold $b \in \{10, 20, \ldots, 1000\}$ kb is tested in combination with each scaling factor $f_2 \in \{1.1, 1.2, \ldots, 2.0\}$.

Training: Each initial SHMM(2) is trained with all emission sequences using the

6. Analysis of Breast Cancer Gene Expression Data

Bayesian Baum-Welch algorithm developed in Sec. 5.3. Again, for comparison reasons, a selected initial $SHMM(2)$ is also trained using the Baum-Welch algorithm developed for the $SHMM(C)$ in Seifert (2006). The training is stopped if the improvement of the optimization function is less than 10^{-9} for two successive training steps.

Detection of differentially expressed genes: Differentially expressed genes are determined like previously described for the *HMM*.

6.2.3 Related approaches from the field of Array-CGH analysis

Closely related to the analysis of tumor expression data in the context of chromosomal locations of genes is the analysis of array comparative genomic hybridization (Array-CGH) data to identify deletions and amplifications of DNA segments in tumor DNA in comparison to DNA from healthy tissue. Generally, the analysis of Array-CGH data is done by considering log-ratios of measured intensities of tumor to reference in the context of the chromosomal locations of the underlying probes along the chromosome (Lai et al. (2005)). The goal is to identify chromosomal regions with log-ratios much greater or much less than zero. Here, chromosomal regions with log-ratios much greater than zero are associated with amplifications and those with log-ratios much less than zero are expected to represent deletions. Unchanged chromosomal regions between tumor and reference are expected to be characterized by log-ratios about zero. For the application of Array-CGH analysis approaches to tumor expression data, chromosomal regions with log-ratios greater than zero are expected to represent over-expressed genes, while under-expressed genes are associated with log-ratios less than zero. This indicates that it might be useful to test such approaches on tumor expression data. Vice versa, the *HMM* developed here has already been demonstrated to work on Array-CGH data (Seifert et al. (2009a)).

In recent years, several methods have been proposed for the analysis of Array-CGH data, and two studies by Lai et al. (2005) and by Willenbrock and Fridlyand (2005) have focused on their comparison to reveal general characteristics of the methods. In addition to this, the ADaCGH web-server has been developed by Diaz-Uriarte and Rueda (2007) to ease the application of different Array-CGH analysis methods and to unify their outputs. All methods listed in Tab. 6.1 except ChARM are available through the ADaCGH web-server, and all these methods including ChARM are subsequently applied with their standard parameter settings to the breast cancer gene expression data set to characterize their ability to identify differentially expressed genes. In analogy to the *HMM* or the *SHMM*(2), the output of all methods of the ADaCGH web-server provides the basics to directly assign one of the three labels '−' (under-expressed), '=' (unchanged expressed), or '+' (over-expressed) to each gene in an experiment. For ChARM each gene that has been reported as not significantly changed is labeled as '=', and each significantly changed gene is labeled either as '−' or '+' in dependency of the sign of the corresponding log-ratio.

Based on Tab. 6.1, the two *HMM*-based methods FHMM and BioHMM are of special interest. The FHMM analyzes the log-ratios in the context of chromosomal locations by partitioning these log-ratios into states that represent the chromosomal aberrations. The BioHMM extends this approach by additionally integrating the chromosomal distance of directly adjacent probes on chromosomes. However, both methods do not integrate prior knowledge into the training. Additionally, also GLAD could be of special interest, because as reported by Lai et al. (2005) also single probes have been identified as changed by this approach. This behavior could be useful for the detection of under-expressed and over-expressed genes that are not included in a greater chromosomal region of under-expressed or over-expressed genes.

Short Name	Method	Reference
ACE	Analysis of Copy Errors	Lingjaerde et al. (2005)
BioHMM	inhomogeneous first-order *HMM*	Marioni et al. (2006)
CBS	Circular Binary Segmentation	Olshen et al. (2004)
CGHseg	CGH segmentation	Picard et al. (2005)
ChARM	Chromosomal Aberration Region Miner	Myers et al. (2004)
GLAD	Gain and Loss Analysis of DNA	Hupé et al. (2004)
FHMM	homogeneous first-order *HMM*	Fridlyand et al. (2004)
Wavelet	Haar wavelet and clustering	Hsu et al. (2005)

Table 6.1: Methods of the Array-CGH data analysis field that are tested to predict differentially expressed genes in the breast cancer data set. All methods except ChARM are provided by the ADaCGH web-server (Diaz-Uriarte and Rueda (2007)).

6.3 Breast Cancer Gene Expression Data Analysis

6.3.1 Comparison of Baum-Welch and Bayesian Baum-Welch training

The Baum-Welch algorithm trains the parameters of the *HMM* and the *SHMM*(2) without including biological prior knowledge about the potential ranges of log-ratios that are

6. Analysis of Breast Cancer Gene Expression Data

expected to represent under-expressed, unchanged expressed, and over-expressed genes in breast cancer. The Bayesian Baum-Welch algorithm integrates this knowledge by the usage of the specified prior. The influence of both training algorithms on the emission parameters of the trained *HMM* and of the trained *SHMM*(2) is shown in Tab. 6.2 and visualized for the *HMM* in Fig. 6.4. The state-specific emission parameters that are obtained from the Baum-Welch algorithm for the states '−' and '+' are clearly different from those that are obtained from the Bayesian Baum-Welch algorithm. Here, the means of the states '−' and '+' obtained from the Baum-Welch algorithm are close to zero, while the corresponding means obtained from the Bayesian Baum-Welch algorithm have a much greater distance to zero. Based on the histogram of log-ratios and the quantile-quantile plot in Fig. 6.1, under-expressed genes modeled by state '−' and over-expressed genes modeled by state '+' are expected to have log-ratios clearly different from zero. Thus, the *HMM* and the *SHMM*(2) that have been trained by the Baum-Welch algorithm have a much lower capability to distinguish under-expressed genes or over-expressed genes from unchanged expressed genes in comparison to the corresponding models obtained through the application of the Bayesian Baum-Welch algorithm. This is even more reflected by the state-specific standard deviations. Here, the ranges of log-ratios that are modeled by the states '−' and '+' clearly overlap with each other for the emission parameters obtained from the Baum-Welch algorithm. This undesired overlap can lead to contradicting predictions of under-expressed or over-expressed genes, which are not reflected in the corresponding log-ratios of the genes. The separation of under-expressed and over-expressed genes is much better reflected in the emission parameters that are obtained from the Bayesian Baum-Welch algorithm. For these reasons, the Bayesian Baum-Welch algorithm should be preferred for the training of the *HMM* and the *SHMM*(2). Consequently, only the models that have been trained by the Bayesian Baum-Welch algorithm are considered in the following sections.

6.3.2 Comparison of HMM, SHMM, and related approaches

All genes predicted as under-expressed or over-expressed by the *HMM*, the *SHMM*(2), or by the related approaches in Tab. 6.1 characterize the potential of each of these methods to identify differentially expressed genes in breast cancer. To analyze this potential for each method, the percentage of predicted under-expressed and over-expressed genes in relation to the total number of genes is considered. In addition to this, the log-ratios of all genes predicted as under-expressed or predicted as over-

6. Analysis of Breast Cancer Gene Expression Data

Model	Training	μ_-	$\mu_=$	μ_+	σ_-	$\sigma_=$	σ_+
HMM	Baum-Welch	-0.04	-0.07	0.23	1.37	0.48	0.60
SHMM(2)	Baum-Welch	-0.03	-0.07	0.23	1.37	0.48	0.60
HMM	Bayesian Baum-Welch	-1.94	-0.01	1.91	1.10	0.55	0.79
SHMM(2)	Bayesian Baum-Welch	-1.93	-0.01	1.90	1.13	0.55	0.82

Table 6.2: Overview of emission parameters of the *HMM* and the *SHMM*(2) obtained from the Baum-Welch algorithm and the Bayesian Baum-Welch algorithm. For each state $i \in S$ the mean μ_i and standard deviation σ_i of the state-specific Gaussian emission density are shown. The Gaussian emission densities of the *HMM* are additionally visualized in Fig. 6.4. For the *HMM* and *SHMM*(2) with $b = 100$ kb and $f_2 = 1.8$, the Baum-Welch algorithm fails to characterize under-expressed genes represented by state '−' and over-expressed genes modeled by state '+'. Such genes are expected to have log-ratios much less or much greater than zero (Fig. 6.1).

expressed are characterized separately by computing the mean, the median and the standard deviation of log-ratios. Here, the percentage of predictions quantifies the prediction behavior of each method, and this measure in combination with the mean, the median, and the standard deviation of the log-ratios allows to analyze the ability to predict differentially expressed genes.

The results of this characterization are summarized in Tab. 6.3. Here, Wavelet, ACE, BioHMM, and FHMM tend to make many more predictions of under-expressed and over-expressed genes than all other methods. However, for these four methods, the mean, the median, and the standard deviation of the corresponding log-ratios show that a large proportion of predicted differentially expressed genes have log-ratios that are close to zero. Additionally, even contradicting predictions have occurred. That is, genes predicted as under-expressed can have log-ratios greater than zero, and vice versa, genes predicted as over-expressed can have log-ratios less than zero. ChARM, CGHseg, and CBS make much less predictions than the four previous methods, but still the same problems occur. In consideration of the histogram of log-ratios and the quantile-quantile plot shown in Fig. 6.1, differentially expressed genes are expected to have log-ratios much less or much greater than zero. In addition to this, the lack of scores for ranking the outputs of the methods of the ADaCGH web-server (Tab. 6.1) makes it difficult to compare these methods. For all these reasons, Wavelet, ACE, BioHMM, FHMM, ChARM, CGHseg, and CBS are not further considered to predict differentially expressed genes in the breast cancer data set.

In contrast to this, GLAD, the *HMM*, and the *SHMM*(2) are the only approaches which predict under-expressed and over-expressed genes in breast cancer like expected from Fig. 6.1. This is reflected by their corresponding values given in Tab. 6.3 for the mean,

6. Analysis of Breast Cancer Gene Expression Data

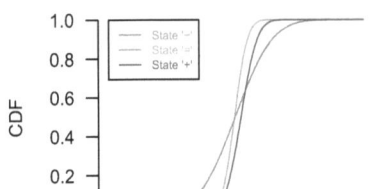
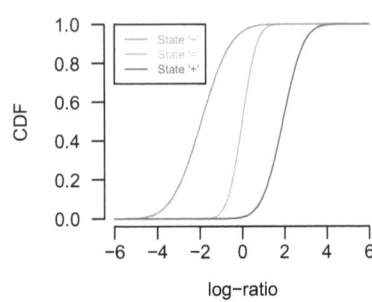

Figure 6.4: Overview of the Gaussian emission densities of the *HMM* obtained by the application of the Baum-Welch algorithm (left) and by the Bayesian Baum-Welch algorithm (right). For each of the three states '−', '=', and '+' of the *HMM* the cumulative distribution function (CDF) of the corresponding Gaussian emission density is shown. The Baum-Welch algorithm clearly fails to characterize under-expressed genes modeled by state '−' and over-expressed genes modeled by state '+'. In contrast to this, the Gaussian emission densities obtained from the Bayesian Baum-Welch algorithm are well separated and characterize under-expressed, unchanged expressed, and over-expressed genes like expected from Fig. 6.1.

the median, and the standard deviation of the log-ratios for under-expressed and over-expressed genes. That means, genes predicted as under-expressed and genes predicted as over-expressed are well separated from each other by their log-ratios. Thus, only GLAD, the *HMM*, and the *SHMM*(2) are considered for further analyses.

6.3.3 Effect of chromosomal distances of genes on self-transition probabilities of SHMMs

The log-ratios measured for two adjacent genes on a chromosome in the breast cancer gene expression data set are clearly depending, like shown in Fig. 6.2, on the chromosomal distance of these genes. That is, log-ratios of adjacent genes in close chromosomal proximity tend to be higher correlated than those of genes in greater distance. The *SHMM*(2) integrates this observation by using a global distance threshold b and a scaling factor f_2 to model that adjacent genes in close chromosomal proximity have a higher probability to be represented by the same state of the *SHMM*(2) than adjacent genes in greater distance. The global distance threshold b is used to assign each pair of two directly adjacent genes on a chromosome either into the group of near gene

6. Analysis of Breast Cancer Gene Expression Data

Method	'−' in %	Mean	Median	Sd	'+' in %	Mean	Median	Sd
Wavelet	18.58	-0.09	-0.09	0.63	9.52	0.13	0.14	0.84
ACE	10.21	-0.56	-0.44	0.80	10.86	0.61	0.53	0.77
BioHMM	7.41	-0.30	-0.23	0.88	9.96	0.39	0.38	0.90
FHMM	6.37	-0.37	-0.27	0.75	5.42	0.62	0.49	0.92
ChARM	1.02	-0.30	-0.27	0.66	1.84	0.31	0.25	0.77
CGHseg	2.45	-0.11	-0.12	0.64	0.97	0.33	0.34	1.10
CBS	2.66	-0.19	-0.17	0.72	1.91	0.47	0.40	0.98
GLAD	1.54	-1.95	-1.74	1.00	1.77	1.85	1.75	0.76
HMM	1.44	-2.25	-2.05	0.90	2.16	1.97	1.86	0.65
SHMM(2)	1.44	-2.22	-2.05	0.95	2.18	1.93	1.85	0.68

Table 6.3: General characterization of genes predicted as under-expressed ('−') and as over-expressed ('+') by the *HMM*, the *SHMM*(2), and the related approaches summarized in Tab. 6.1. The percentages of genes predicted as '−' and '+', and the corresponding mean, median, and standard deviation of the log-ratios are shown for each method. For the *HMM* and the *SHMM*(2) with global distance threshold $b = 100$ kb and scaling factor $f_2 = 1.8$ the Viterbi algorithm has been used for the predictions.

pairs or into the group of far gene pairs. The scaling factor f_2 allows to set the degree to which two adjacent genes of the near gene pair group tend to be modeled by the same state of the *SHMM*(2). To analyze the effect of modeling chromosomal distances of adjacent genes by the *SHMM*(2), the influence of different parameter combinations of b and f_2 on the self-transition probabilities of the *SHMM*(2) in Fig. 6.3 is evaluated. The potential that two genes of the near or of the far gene pair group are represented by the same state of the model is directly characterized by the self-transition probability of this state. To investigate different combinations of b and f_2 systematically, all values of b from 10 kb up to 1000 kb in steps of 10 kb have been tested in combination with each scaling factor f_2 from 1.0 to 2.0 in steps of 0.1 by training the corresponding *SHMM*(2) with the Bayesian Baum-Welch algorithm. The obtained behavior of the self-transition probabilities is summarized in Fig. 6.5 for a representative selection of models. Generally, two trends for each single selection can be clearly observed.

1. The self-transition probabilities for the groups of near and far gene pairs decreases for a fixed scaling factor f_2 and an increasing distance threshold b (Fig. 6.5a).

2. The self-transition probabilities for the group of far gene pairs decrease while those for the group of near gene pairs increase for a fixed distance threshold b and an increasing scaling factor f_2 (Fig. 6.5b).

The first trend reflects the observation shown in Fig. 6.2 that the correlation of the log-ratios measured for adjacent gene pairs tends to decrease with increasing chromosomal gene distance. Thus, for a fixed scaling factor it is more unlikely that two adjacent genes are represented by the same state of the *SHMM*(2) for an increasing distance threshold. The second trend shows that by increasing the scaling factor the genes in the near gene pair group have a higher probability to be modeled by the same state than genes in the far gene pair group. How these trends influence the ability of the *SHMM*(2) to predict differentially expressed genes in breast cancer is investigated in the following section.

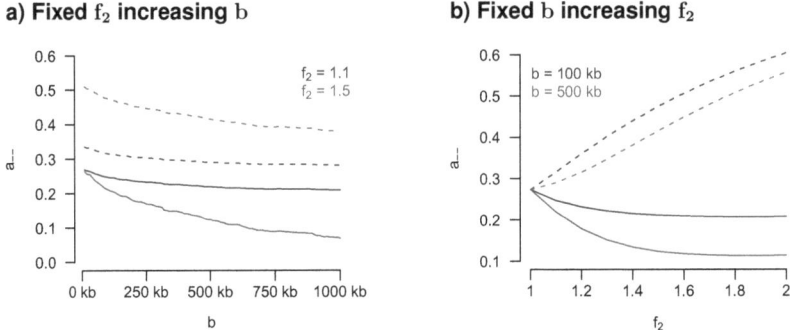

Figure 6.5: Behavior of the self-transition probability 'a_ _ _' of the state '−' of the *SHMM*(2) in Fig. 6.3 for different combinations of the scaling factor f_2 and the global distance threshold b. Self-transition probabilities for the group of near gene pairs with genes in distance less or equal than b are represented by dashed lines. For the group of far gene pairs with genes in distance greater than b the corresponding self-transition probabilities are shown by solid lines. The value of the self-transition probability of the state '−' of the *HMM* that does not model chromosomal distances of genes is shown in the right figure for $f_2 = 1$.

6.3.4 Validation of prediction results of HMM, SHMMs, and GLAD

Amplifications and deletions of DNA segments are known to have direct effects on the expression levels of the affected genes in breast cancer (Hyman et al. (2002); Pollack et al. (2002)). Generally, amplified genes tend to be highly expressed and deleted genes are associated with lower expression levels in comparison to a normal reference sample. For the breast cancer gene expression data set also amplifications and deletions of genes have been measured by Pollack et al. (2002) for each cell line and each

6. Analysis of Breast Cancer Gene Expression Data

tumor using the Array-CGH approach. The direct effects of amplifications and deletions on the corresponding gene expression levels are clearly shown in Fig. 6.6. Most of the amplified genes are associated with gene expression log-ratios much greater than zero as expected for over-expressed genes in Fig. 6.1. The direct effects of deletions on the gene expression levels are also present for a greater fraction of deleted genes which tend to be under-expressed based on their gene expression log-ratios that are much less than zero. Due to the lack of information about differentially expressed genes and the generally known diversity of gene expression profiles of individual breast tumors (Perou et al. (2000)), the usage of information about individual deletions and amplifications of genes can be considered for the comparison of the *HMM*, the $SHMM(2)$, and GLAD. That means, these methods should be able to predict deleted genes as under-expressed and amplified genes as over-expressed. This does of course not fully hold for each gene in the breast cancer gene expression data set, because genes can be under-expressed or over-expressed without any underlying deletion or amplification. However, this strategy makes use of the only individual validation data that is available for the breast cancer gene expression data set.

To validate the three methods, the breast cancer Array-CGH data set by Pollack et al. (2002) has been used to label each gene in the 4 breast cancer cell lines and the 37 breast tumors of the corresponding breast cancer gene expression data set. In analogy to Pollack et al. (2002), a gene has been labeled as deleted if its Array-CGH log-ratio is less than -1.585, as amplified if its log-ratio is greater than 1.585, and otherwise labeled as unchanged. In total, 228,677 genes have been labeled as unchanged. A small fraction of 125 genes has been labeled as deleted and 228 genes have been labeled as amplified. Together, deleted and amplified genes form the positive class containing genes that should be predicted as under-expressed or as over-expressed based on their characteristics of log-ratios shown in Fig. 6.6. The negative class consists of the genes that have been labeled as unchanged. The majority of genes in this class should be predicted as unchanged expressed, because like shown in Fig. 6.6 most of the corresponding log-ratios have values about zero. In addition to this, it is also expected that a greater fraction of genes from the negative class is predicted as under-expressed or over-expressed. This occurs due to the fact that deletions and amplifications of DNA segments are not the only causes that lead to differential expression in breast cancer.

Initially, the performance of GLAD is investigated. Based on the initial study summarized in Tab. 6.3, GLAD is able to predict over-expressed and under-expressed genes

6. Analysis of Breast Cancer Gene Expression Data

that have log-ratios as expected from Fig. 6.1, but GLAD does not provide scores to rank the predictions. For that reason, the point measures of true positive rate (TPR) and false positive rate (FPR) have been computed for the predictions obtained by GLAD for the breast cancer gene expression data set. GLAD identifies 26.98% (TPR) of the amplified genes as over-expressed and deleted genes as under-expressed at a FPR of 3.39%. Next, to compare the predictions of the *HMM* and the *SHMM*(2) against the predictions of GLAD, the TPRs of the *HMM* and the *SHMM*(2) have been evaluated at the fixed FPR of 3.39% given by GLAD. An overview of the results is shown in Fig. 6.7. At the level of 3.39% FPR the *HMM* predicts 37.68% of the amplified genes as over-expressed and deleted genes as under-expressed. The best *SHMM*(2) models predict 39.94% of these genes. This is clearly more than obtained by GLAD and the *HMM*. The large fraction of *SHMM*(2) models that are better than the *HMM* is shown in Fig. 6.7 colored in yellow, and the small proportion of models that have a smaller TPR than the *HMM* is colored in blue.

In summary, the direct effects of amplifications and deletions of DNA segments on the expression levels of the affected genes shown in Fig. 6.6 can be predicted best as represented in Fig. 6.7 by the *SHMM*(2) in comparison to the *HMM* and GLAD. Subsequently, the influence of modeling chromosomal locations and distances of genes on the prediction of under-expressed and over-expressed genes is investigated in more detail.

6.3.5 Influence of modeling chromosomal locations and distances of genes on the prediction results

For the breast cancer gene expression data set the gene expression levels of directly adjacent genes on chromosomes are observed to be positively correlated as shown in Fig. 6.2. Beyond that, it can be clearly seen that the gene expression levels of directly adjacent genes in close chromosomal proximity tend to be higher correlated than those of adjacent genes in greater distance. Here, the goal is to investigate how the modeling of these observations influences the prediction of under-expressed and over-expressed genes in breast cancer.

The mixture model (e.g. Bilmes (1998)) consisting of three Gaussian densities for modeling under-expressed, unchanged expressed, and over-expressed genes can be considered as basic model for analyzing the breast cancer data set. This model neither integrates chromosomal locations nor chromosomal distances of genes into the pre-

6. Analysis of Breast Cancer Gene Expression Data

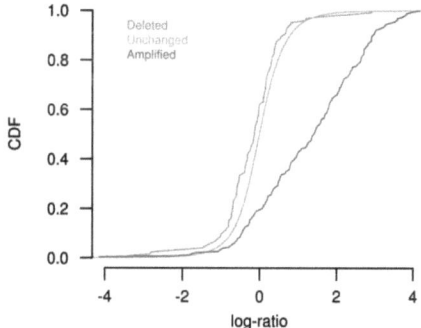

Figure 6.6: Overview of gene expression levels in the breast cancer gene expression data set in the context of the underlying DNA status of the genes. Genes have been group into the three categories deleted, unchanged, and amplified according to the Array-CGH data provided by Pollack et al. (2002). In analogy to Pollack et al. (2002), a gene is labeled as deleted if its Array-CGH log-ratio is less than -1.585, as amplified if its log-ratio is greater than 1.585, and otherwise labeled as unchanged. For each category, the cumulative distribution function is shown for the corresponding gene expression log-ratios of all genes in this category. Direct effects of deletions and amplifications on the gene expression level are observed in comparison to the unchanged category.

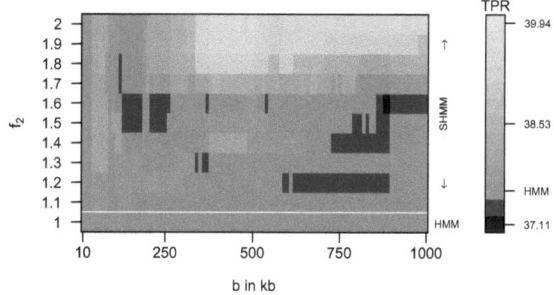

Figure 6.7: TPRs obtained for the *HMM* and the *SHMM*(2) for the prediction of the direct effects of amplifications and deletions on the gene expression levels in the breast cancer data set (Fig. 6.6) at the level of the FPR of 3.39% of GLAD. Most of the *SHMM*(2) models perform better than the *HMM*. Both types of models are better than GLAD which has a TPR of 26.98%. ROC curves of the *HMM* and of one of the best *SHMM*(2) models are shown in Fig. 6.8.

97

diction of under-expressed or over-expressed genes. In contrast to this, the *HMM* considers chromosomal locations of genes by modeling dependencies between directly adjacent genes on a chromosome, and the *SHMM*(2) additionally integrates the chromosomal distance of directly adjacent genes. To compare these three models, a mixture model with the same initial and prior settings like specified for the *HMM* and the *SHMM*(2) has additionally been trained on the breast cancer data set. The resulting mixture model represents under-expressed genes by a Gaussian density with mean -1.96 and standard deviation 1.04, unchanged expressed genes are characterized by a Gaussian density with mean 0.01 and standard deviation 0.56, and a Gaussian density with mean 1.95 and standard deviation 0.73 models the over-expressed genes. Like motivated in Fig. 6.1, most of the genes are expected to be unchanged expressed. This is reflected in the mixture model by a mixture weight of 0.96 for the Gaussian density that models unchanged expressed genes, while under-expressed and over-expressed genes are represented by weighting the corresponding Gaussian density by 0.02. The parameters obtained for the Gaussian densities of the mixture model are very similar to those obtained for the *HMM* and the *SHMM*(2) shown in Tab. 6.2. This is not surprising since the *HMM* and the *SHMM*(2) are specific extensions of the mixture model.

In analogy to the previous section, the validation data shown in Fig. 6.6 is used for the comparison of GLAD, the mixture model, the *HMM*, and the *SHMM*(2). The receiver operating characteristic (ROC) curves obtained for the mixture model, the *HMM*, and one of the best *SHMM*(2) models in Fig. 6.7 are shown in Fig. 6.8 including the point measures of TPR and FPR obtained for GLAD. All three models are much better than GLAD that analyzes the log-ratios in the contexts of chromosomal locations of genes. In addition to this, one clearly observes that the modeling of chromosomal locations of genes by the *HMM* leads to an improved prediction of under-expressed and over-expressed genes in comparison to the mixture model that ignores these information. This prediction performance is further improved by the *SHMM*(2) that additionally integrates distances of directly adjacent genes on chromosomes. Thus, the direct effects of amplifications and deletions on the gene expression levels shown in Fig. 6.6 are predicted best by modeling chromosomal distances of genes by the *SHMM*(2).

6.3.6 Hotspots of under-expression and over-expression

Individual breast tumors are known to have very diverse gene expression profiles (Perou et al. (2000)). Publicly available databases like the Genetic Association Database (GAD) by Becker et al. (2004) or the Breast Cancer Database (BCD) by

6. Analysis of Breast Cancer Gene Expression Data

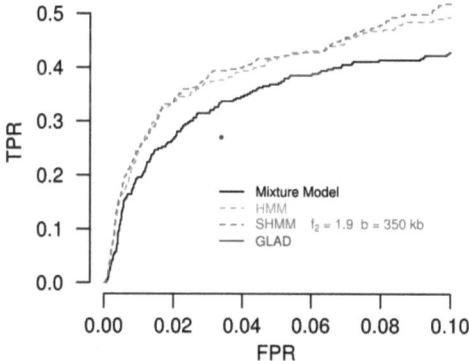

Figure 6.8: Important parts of the ROC curves of the mixture model, the *HMM*, and the *SHMM*(2) for the prediction of amplified genes as over-expressed and of deleted genes as under-expressed in the breast cancer gene expression data set. The point measures of TPR and FPR obtained for GLAD are represented by the blue dot. The *SHMM*(2) is among the best models in Fig. 6.7.

Telikicherla et al. (2008) collect genes that have been identified to play a role in breast cancer. The GAD contains 67 genes and the BCD has 1361 genes that have also been measured in the breast cancer gene expression data set by Pollack et al. (2002). The overlap of these genes in both databases comprises 47 genes. However, the individuality of breast tumors does not allow to use the genes contained in these databases as candidate genes of under-expression or over-expression for each individual gene expression profile in the breast cancer gene expression data set. Anyhow, genes that are frequently predicted as under-expressed or over-expressed in the breast cancer data set can be compared to these two databases for identifying those candidate genes that are not contained in both databases. These candidate genes can be further investigated for their role in breast cancer by additional literature studies.

To investigate this, the *SHMM*(2) with scaling factor $f_2 = 1.9$ and distance threshold $b = 350$ kb is used to predict under-expressed and over-expressed genes in the breast cancer gene expression data set at the level of the FPR of GLAD. This *SHMM*(2) has been shown in Fig. 6.7 and Fig. 6.8 to be a useful tool for the prediction of direct effects of amplifications and deletions on the gene expression levels in breast cancer. Based on that, each gene which has been predicted at least 7 times as under-expressed (or

6. Analysis of Breast Cancer Gene Expression Data

as over-expressed) by this $SHMM(2)$ in the 41 breast cancer expression profiles has been further analyzed. In total, 49 genes that fulfill this criterion have been identified, and 33 of these genes are already contained in the GAD or in the BCD. The remaining 16 genes given in Tab. 6.4 have been further investigated by additional literature searches, whereas 13 of these genes could be directly associated with breast cancer. Of the three remaining genes, AA088457 and CBX2 have been found to play a role in other types of cancer, and CYP4X1 has been observed as over-expressed in breast cancer in its NCBI Unigene EST profile.

In summary, this analysis shows that the $SHMM(2)$ also identifies genes known to play a role in breast cancer that are currently not included in the two public databases GAD and BCD. This provides further support that the $SHMM(2)$ can be used for the prediction of differentially expressed genes in breast cancer. In addition to this, the $SHMM(2)$ might also be applied to the analysis of other tumor data sets.

Gene	Annotation	Prediction	Literature Search
AA088457	tumor specific mitosis dependent	$+$	Xu et al. (2007)
ADH2	alcohol dehydrogenase subunit beta	$+$	Perou et al. (2000)
CBX2	cell division cycle associated	$+$	Raaphorst (2005)
CEACAM5	carcinoembryonic antigen	$-/+$	Blumenthal et al. (2007)
COL11A1	collagen, alpha 1	$-$	Halsted et al. (2008)
CYP4X1	cytochrome P450 family	$-/+$	NCBI UniGene EST profile
CYP4Z1	cytochrome P450 family	$-/+$	Rieger et al. (2004)
DLX4	distal-less homeobox 4	$+$	Tomida et al. (2007)
H11	cell proliferation and apoptosis	$+$	Depre et al. (2002)
HE4	protease inhibitor	$-/+$	Galgano et al. (2006)
HLA-DQA2	major histocompatibility complex	$-$	Maiorana et al. (1995)
IGL@	immunoglobulin lambda locus	$-$	Yu et al. (2004)
MGB1	mammaglobin	$-$	Sasaki et al. (2007)
MLN64	metastatic lymph node protein	$+$	Kauraniemi et al. (2001)
SCYB14	chemokine in breast and kidney	$-$	Ma et al. (2007)
UGT2B7	catechol estrogen specific	$-$	Gestl et al. (2002)

Table 6.4: Genes predicted at least seven times as under-expressed '$-$', over-expressed '$+$', or both '$-/+$' by the $SHMM(2)$ with scaling factor $f_2 = 1.9$ and distance threshold $b = 350$ kb in the breast cancer gene expression data set. None of these genes is contained in the GAD or in the BCD. The GeneCards database (http://www.genecards.org) has been used to annotate the genes. All genes except AA088457, CBX2, and CYP4X1 are directly associated with breast cancer based on the specified literature. The genes AA088457 and CBX2 have support from other types of cancer, and CYP4X1 has been observed as over-expressed in breast cancer based on its NCBI UniGene EST profile (http://www.ncbi.nlm.nih.gov).

6.4 Further reading

Selected parts and some additional extensions of this study have recently been published (Seifert et al. (2011)). A conceptual extension of the *SHMM* by replacing the fixed transition matrices by a continuous transition function of gene distance has been made. This has led to a Hidden Markov Model with distance-scaled transition matrices (*DSHMM*) directly integrating the distance of adjacent genes into the state-transition process. Such a model avoids potential discretization effects caused by non-continuous changes of transition probabilities for small variations in gene distances. *SHMM* and *DSHMM* performed nearly identical in an in-depth comparison study. However, the integration of prior knowledge into the training is more important than the additional modeling of chromosomal distances of genes. A Java-based implementation of mixture models, *HMMs*, *SHMMs* and *DSHMMs* for analyzing tumor gene expression data is publicly available from http://www.jstacs.de/index.php/DSHMM.

7 Analysis of Promoter Array ChIP-chip Data

In recent years array-based analysis of chromatin immunoprecipitation (ChIP-chip) data has become a powerful technique to identify DNA target regions of individual transcription factors. ChIP-chip has firstly been applied to yeast by Ren et al. (2000) and Iyer et al. (2001) based on promoter arrays. With the availability of sequenced genomes, ChIP-chip is predominantly based on tiling arrays that represent a genome by DNA probes at high resolution (Johnson et al. (2008)). The analysis of ChIP-chip data is challenging, because of the huge data sets containing thousands of hybridization signals. Most available methods focus on the analysis of tiling array ChIP-chip data to predict chromosomal target regions of DNA-binding proteins like transcription factors or histones. Examples include a moving average method by Keles et al. (2004), an *HMM* approach by Li et al. (2005), TileMap by Ji and Wong (2005) using moving averages or an *HMM* to account for information about adjacent probes, or PMT by Chung et al. (2007) that integrates a physical model to correct for probe-specific behavior. More recently, a new *HMM* approach has been developed by Humburg et al. (2008), outperforming TileMap in the context of the prediction of histone modifications from tiling array ChIP-chip data. Also ChIPmix (Martin-Magniette et al. (2008)) that utilizes a linear regression mixture model is appropriate for this analysis.

In this chapter, the focus is on the development of *HMM*-based methods for the analysis of promoter array ChIP-chip data. Promoters are functional parts of the DNA that are typically located upstream in close chromosomal proximity of genes (Davidson (2001)). Specific DNA sequences within promoters are recognized by transcription factors to regulate the expression of genes (Davidson (2001); Latchman (2004)). Promoters of genes can be represented on a promoter array to identify target genes of a specific transcription factor in a ChIP-chip experiment. Two directly adjacent genes on a chromosome can be located in head-head, tail-tail, tail-head, or head-tail orientation to each other. This is illustrated in the upper part of Fig. 7.1. The head-head orientation is of special interest because a promoter array can represent both genes separately by their corresponding promoter fragments. This can lead to overlapping promoter fragments if both genes are in close chromosomal distance to each other. Thus, also in dependence of the length of the DNA segments that are hybridized to the promoter array, it is expected that the measurements for gene pairs in head-head orientation are more similar to each other than the measurements of gene pairs in other orientations. This trend is clearly observed for the promoter array ChIP-chip data of the seed-specific transcription factor ABI3 of the model plant *Arabidopsis thaliana* (Fig. 7.1) and for the ChIP-chip data of the cell cycle specific transcription factors ACE2, SWI5, and FKH2 of the yeast *Saccharomyces cerevisiae* (Tab. 7.1). Motivated by this observation, these promoter array ChIP-chip data sets are analyzed in the context of chromosomal locations of genes by an *HMM* following the two-state architecture proposed by Li et al. (2005) for the analysis of ChIP-chip tiling array data. In addition to this, the *HMM* is extended to a $SHMM(2)$ with two scaled transition matrices that specifically models gene pairs in head-head orientation. Both approaches are compared to the standard log-fold change analysis to investigate whether the integration of chromosomal locations and information about gene pair orientations helps to improve the prediction of transcription factor target genes. Predicted target genes are validated using literature and database searches, publicly available gene expression data, and independent wet-lab experiments.

Goals of this Chapter

1. The promoter array ChIP-chip data sets of *S. cerevisiae* and *A. thaliana* are introduced.

2. The standard log-fold change analysis, the *HMM*, and the $SHMM(2)$ for the analysis of promoter array ChIP-chip data are described.

3. These three approaches are initially tested on the ChIP-chip data of *S. cerevisiae* and comprehensively studied on the data set of *A. thaliana*.

7. Analysis of Promoter Array ChIP-chip Data

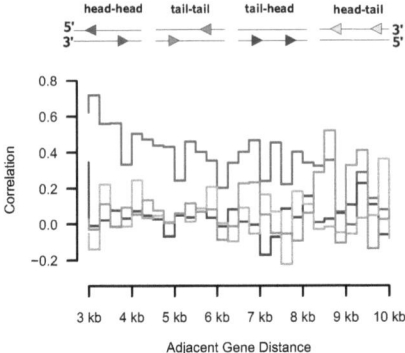

Figure 7.1: Pearson correlations of promoter array ChIP-chip measurements of the transcription factor ABI3 in the context of the four gene pair orientations head-head, tail-tail, tail-head, and head-tail of two directly adjacent genes on DNA in distances of 3 kb up to 10 kb in steps of 250 bp. Genes are represented by triangles, and the orientation of the tip of a triangle defines the reading direction of a gene. The promoter fragment of a gene in the ABI3 data set is always located in 3' direction of the gene. The ChIP-chip measurement of a gene is the $\log_2$-ratio of immunoprecipitated DNA for ABI3 to input control DNA that is measured for the corresponding promoter of the gene. The intergenic region between two genes in head-head orientation is represented by two promoter fragments, one for each gene. Depending on the distance between these two genes the extracted DNA segments in the immunoprecipitated sample and in the input DNA sample can bind to both promoter fragments of these two head-head genes leading to significantly higher correlations for genes in head-head orientation in comparison to all other gene pair orientations.

TF	head-head	tail-tail	tail-head	head-tail
ACE2	0.76	0.37	0.13	0.26
SWI5	0.80	0.26	0.12	0.20
FKH2	0.89	0.29	0.27	0.22

Table 7.1: Pearson correlations of promoter array ChIP-chip measurements of the transcription factors ACE2, SWI5, and FKH2 for the four gene pair orientations head-head, tail-tail, tail-head, and head-tail based on all pairs of two directly adjacent genes in the data set by Lee et al. (2002). The correlations of ChIP-chip measurements of gene pairs in head-head orientation are clearly higher than in the three other categories.

7. Analysis of Promoter Array ChIP-chip Data

7.1 Promoter Array Data Sets

7.1.1 Yeast Data Set

Publicly available promoter array ChIP-chip data from Lee et al. (2002) provides the basics to identify common target genes of the cell cycle specific transcription factors ACE2 and SWI5, and ACE2 and FKH2. First, all ChIP-chip measurements have been mapped to their corresponding positions in the genome of *S. cerevisiae* using the Saccharomyces Genome Database by Cherry et al. (1997). Then, the log-ratio $o_t(k)$ of immunoprecipitated DNA to input DNA has been computed for each gene t on chromosome k based on its corresponding ratio which has been measured in the ChIP-chip experiment for the promoter of each gene. Here, for each of the three considered transcription factors one emission sequence $\vec{o}(k) = (o_1(k), \ldots, o_{T_k}(k))$ is obtained for the k-th chromosome of the in total sixteen chromosomes of *S. cerevisiae*. The log-ratios in $\vec{o}(k)$ are systematically ordered from the left arm to the right arm of chromosome k.

7.1.2 Arabidopsis Data Set

Promoter array ChIP-chip data of the seed-specific transcription factor ABI3 has been generated at the IPK Gatersleben during the project Arabido-Seed (2004-2009) based on seeds of *A. thaliana* accession Columbia (Col). This inhouse data set provides the opportunity to identify target genes of ABI3 for gaining more detailed insights into seed development based on 11,904 promoters of genes represented on a promoter array. Each ABI3 ChIP-chip experiment comprises two sub-experiments on separate promoter arrays. In the first sub-experiment, enriched DNA fragments bound by ABI3 have been measured. The second sub-experiment has been used to measure genomic input DNA of the corresponding *A. thaliana* seeds to obtain reference measurements for differentiating between DNA fragments specifically bound by ABI3 and unspecific bindings. Subsequently, five ABI3 experiments involving three biological replicates whereas two of them have additionally been repeated in a technical replicate are considered. To normalize these five experiments, the sub-experiments of each experiment have firstly been treated separately by shifting the median of the measured $\log_2$-intensities to zero to correct for differences in the absolute signal intensities. In the final normalization step, quantile normalization by Bolstad et al. (2003) has been applied to the five sub-experiments that measure the DNA segments bound by ABI3 and to the five corresponding reference experiments. Next, the log-ratio $o_t(k)$ of DNA bound by ABI3 in relation to genomic input DNA has been computed for each gene t on chromosome k for each of the five experiments using the normalized log$_2$-intensities of the corresponding promoter. Finally, the TAIR7 genome annotation has been used to map all log-ratios of an experiment to their positions in the genome of *A. thaliana*. This leads for each experiment to one emission sequence $\vec{o}(k) = (o_1(k), \ldots, o_{T_k}(k))$ for each chromosome of the five chromosomes of *A. thaliana*.

7.2 Methods for Promoter Array Data Analysis

7.2.1 Standard Log-Fold-Change analysis

The log-ratio of immunoprecipitated DNA to input DNA measured for a promoter characterizes the potential of the corresponding gene to be a target gene of the analyzed transcription factor. Thus, it is expected that putative target genes have log-ratios that are much greater than zero. To identify putative target genes, the standard log-fold-change analysis (*LFC*) is used. *LFC* initially sorts all genes of an experiment in decreasing order of their corresponding log-ratios to represent putative target genes at the top of the resulting list. Then, the sorted lists of all experiments are used to determine the intersection of top candidates of each of these lists. All genes in the intersection are interpreted as putative target genes of the analyzed transcription factor.

7.2.2 Basic first-order Hidden Markov Model

Model: A first-order *HMM* with two states $S := \{-, +\}$ characterized by state-specific Gaussian emission densities is used to analyze the ChIP-chip promoter array data. This *HMM* is a special case of the *HHMM*(L, C) defined in Sec. 3.2 by setting the order to $L = 1$ and by using $C = 1$ transition class. The *MM* that underlies this *HMM* is shown in Fig. 2.2. The state '$-$' of the *HMM* models putative non-target genes of the analyzed transcription factor, and putative target genes are represented by state '$+$'.

Initialization: In general, the initial *HMM* should distinguish putative target genes of the analyzed transcription factor from non-target genes with respect to their log-ratios in the emission sequences. Here, a histogram of log-ratios helps to find good initial *HMM* parameters. The choice of initial parameters addresses the presumptions that the proportion of non-target genes is much greater than that of target genes, and that the number of successive non-target genes is also much greater than the number of successive target genes. To integrate these presumptions, the start probabilities are set to $\pi_- = 0.9$ and $\pi_+ = 0.1$, and the initial transition matrix A_1 is chosen to have the equilibrium distribution $\vec{\pi} = (0.9, 0.1)$ by setting $a_{ii}(1) = 1 - s/\pi_i$ and $a_{ij}(1) = 1 - a_{ii}(1)$ for each $i \in S$ and each $j \in S$ with $j \neq i$ based on $s = 0.05$. This leads to an initial state duration of eighteen for state '−', which is nine times greater than that of state '+'. Finally, the states are characterized by specific means $\mu_- = 0$ and $\mu_+ = 2$, and by proper standard deviations $\sigma_- = 1$ and $\sigma_+ = 0.5$.

Training: The initial *HMM* is trained with all emission sequences using the Bayesian Baum-Welch algorithm developed in Sec. 3.7.1. Biological prior knowledge is included into the training based on the prior specified in Sec. 3.6. That is, $\vartheta_i = 2$ is used for all $i \in S$ as parameter of the start prior, $\vartheta_{ij}(1) = 1$ is used for all $i \in S$ and for all $j \in S$ as parameter of the transition prior, and the parameters of the emission prior are set to $\eta_- = 0$, $\eta_+ = 2$, $\epsilon_- = \epsilon_+ = 10^3$, $r_- = 1$, $r_+ = 100$, and $\alpha_- = \alpha_+ = 10^{-4}$. The choice of this prior ensures a good characterization of both states to distinguish putative target genes from non-target genes. The training is stopped if the increase of the log-posterior of two successive iterations is less than 10^{-9}.

Target gene detection: The state '+' of the trained *HMM* models the potential of genes to be targets of the analyzed transcription factor. Hence, each gene t on each chromosome k in an experiment is considered by computing the state-posterior $\gamma_t^k('+')$ defined in (3.6) to obtain the probability that the underlying gene is a putative target gene. Then, each experiment is analyzed separately by creating a sorted list containing all genes of the experiment in decreasing order of their state-posteriors. Finally, all experiments are used to determine the intersection of the top candidate genes of each sorted list. In analogy to the *LFC* approach, all genes in the intersection are interpreted as putative target genes of the analyzed transcription factor.

7.2.3 Hidden Markov Model with two scaled transition matrices

Model: A *SHMM*(2) with two states $S := \{-, +\}$ characterized by state-specific Gaussian emission densities is used to analyze ChIP-chip data in the context of gene pair orientations on chromosomes. The two-state architecture of the *SHMM*(2) is shown in Fig. 7.2. The *SHMM*(2) is defined in a general form in Sec. 5.2, and the underlying inhomogeneous *MM* with two states and two transition classes is shown in Fig. 2.3. As motivated through Fig. 7.1 and Tab. 7.1, two directly adjacent genes on a chromosome that are located in head-head orientation to each other tend to have log-ratios that are highly positively correlated, whereas log-ratios of other gene pairs tend to be uncorrelated or only weakly positively correlated. This observation is integrated into the data analysis by assuming that it is more likely for two genes in head-head orientation to be represented by the same state of the *SHMM*(2) than for other gene pair orientations. Two scaled transition matrices A_1 and A_2 are used to model this. Both transition matrices are computed on the basis of the basic transition matrix A, which is identical to the previously defined A_1 of the *HMM*, the pre-defined scaling factor $f_1 := 1$, and the user-defined scaling factor $f_2 > f_1$. To specify the corresponding transition matrix for each gene pair, each pair of successive genes t and $t+1$ on each chromosome k is assigned to its corresponding transition class

$$c_t(k) := \begin{cases} 2, & \text{genes } t \text{ and } t+1 \text{ are in head-head and } d_t \leq b \\ 1, & \text{otherwise} \end{cases}$$

in dependence of the category of the considered gene pair, the chromosomal distance d_t of both genes, and the globally defined distance threshold $b \in \mathbb{N}$. That is, a transition from the state of gene t to the state of gene $t+1$ is done by using the corresponding transition matrix $A_{c_t(k)}$. The self-transition probability of each state $i \in S$ increases strictly from transition class A_1 to A_2. Thus, its more likely for a head-head gene pair modeled by A_2 that both genes are represented by the same state of the *SHMM*(2) than for other gene pairs modeled by A_1.

Figure 7.2: The basic two-state architecture of the *SHMM*(2) that is used to analyze promoter array ChIP-chip data with respect to the gene pair orientations on a chromosome. The states $S := \{-, +\}$ are represented by labeled circles and corresponding state-specific Gaussian emission densities. Non-target genes of a transcription factor are modeled by state '$-$' and target genes are represented by state '$+$'. Thick arrows represent transitions for genes in head-head orientation and thin arrows those of genes in other orientations. The gene pair orientations are illustrated in Fig. 7.1.

Initialization: The basic initialization of the *SHMM*(2) is identical to those of the *HMM*. In addition, the distance threshold b for the selection of the transition classes must be defined, and a scaling factor f_2 has to be chosen to specify the degree of differentiation between head-head orientations modeled by A_2 and all other orientations modeled by A_1. Since both values are data set dependent they are specified in the corresponding analyses sections.

Training: The $SHMM(2)$ is trained using the Bayesian Baum-Welch algorithm developed in Sec. 5.3 with prior parameters identical to those defined for the HMM.

Target gene detection: Putative target genes are determined in analogy to the HMM approach.

7.3 Identification of Common Target Genes of Yeast Cell Cycle Regulators

Common putative target genes of the *S. cerevisiae* cell cycle regulators ACE2 and SWI5, and ACE2 and FKH2 are determined based on the predictions of the *LFC*, the *HMM*, and the $SHMM(2)$. The transcription factors ACE2 and SWI5 are known to regulate common target genes expressed at the boundary of the M/G1 phase of the cell cycle (Mc Bride et al. (1999); Lee et al. (2002)), and the transcription factors ACE2 and FKH2 control the regulation of a common set of genes in the G1 phase of the cell cycle (Lee et al. (2002)). Here, the prediction results of the three methods are compared against each other using the Saccharomyces Genome Database (Cherry et al. (1997)) to analyze whether the putative target genes are known to play a role in the regulation of the cell cycle.

7.3.1 Prediction of putative common target genes

The *LFC* method is used to predict putative common target genes of ACE2 and SWI5 at the level of the top 75 candidate genes of each transcription factor, and for ACE2 and FKH2 the top 130 candidate genes of each transcription factor are used. The specified numbers of top candidates ensure that each candidate gene of a transcription factor has a log-ratio greater than one. The application of the *HMM* and the $SHMM(2)$ is motivated by Tab. 7.1. Here, positive correlations between the measured log-ratios of genes in head-head, tail-tail, tail-head, and head-tail orientation have been observed for all transcription factors. This motivates the usage of the *HMM* to model dependencies between directly adjacent log-ratios on a chromosome. Additionally, the correlations in the head-head category are much greater than the correlations observed for the other categories. Based on this, the $SHMM(2)$ is applied to specifically model the gene pairs in head-head orientation. Here, the global distance threshold b required for

7. Analysis of Promoter Array ChIP-chip Data

the specification of the transition classes is set to infinity, because almost all genes in head-head orientation have distances less than 2 kb in combination with the fact that the correlations of the log-ratios of these genes are very high for all three transcription factors. In addition to this, the SHMM(2) has been tested by using the scaling factor $f_2 = 2.0$, $f_2 = 4.0$, and $f_2 = 6.0$. Like specified for LFC, the same numbers of top candidates are considered by the HMM and by the SHMM(2) to predict common target genes of the transcription factors. The SHMM(2) with scaling factor $f_2 = 2.0$ has predicted one putative common target gene less for ACE2 and SWI5 than the SHMM(2) with $f_2 = 4.0$. Besides this, the prediction results have been identical for $f_2 = 4.0$ and $f_2 = 6.0$. For ACE2 and FKH2, each SHMM(2) has identified the same putative common target genes. Due to that, the focus is subsequently on the SHMM(2) with scaling factor $f_2 = 4.0$. The results for the comparison of the LFC, the HMM, and the SHMM(2) are shown in Fig. 7.3. In both cases, all common target genes predicted by the LFC and the HMM have also been predicted by the SHMM(2). Moreover, the SHMM(2) has predicted two putative target genes that have not been identified by the LFC and the HMM. Subsequently, the putative target genes are further investigated based on literature searches.

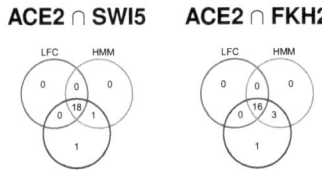

Figure 7.3: Venn diagrams for comparing the prediction of common target genes of the S. cerevisiae cell cycle transcription factors ACE2 and SWI5, and ACE2 and FKH2 by the three methods LFC, HMM, and the SHMM(2) with scaling factor $f_2 = 4.0$ that specifically models the head-head orientation of adjacent gene pairs. In both cases, the SHMM(2) is the most general model that predicted the greatest number of putative common target genes including all target genes predicted by the LFC and the HMM.

7.3.2 Validation of putative common target genes

The Saccharomyces Genome Database (Cherry et al. (1997)) is used to investigate whether the putative common target genes that have only been predicted by the SHMM(2), or together by the HMM and the SHMM(2) are involved in the regulation of the yeast cell cycle. Regarding the common target genes of ACE2 and SWI5, the gene YJL160C has only been predicted by the *SHMM*(2). This gene is a member of the PIR family of cell wall proteins with functions in sporulation, and its gene expression level is weakly cell cycle regulated peaking in the M phase of the cell cycle (Jung and Levin (1999); Giaver et al. (2002); Enyenihi and Saunders (2003); de Lichtenberg et al. (2005)). Currently, no function is known for the gene YBR157C predicted by the *SHMM*(2) and the *HMM*. Considering the common target genes of ACE2 and FKH2, the gene YER127W has only been predicted by the *SHMM*(2). This gene encodes a protein which is essential for the maturation of the 18S rRNA. The repression of the gene expression of this gene leads to an abnormal progression of the G1 phase of the cell cycle (Yu et al. (2006)). The genes YER126C, YFL021W, and YFL022C have been identified as putative common target genes of ACE2 and FKH2 by the *HMM* and by the *SHMM*(2). The protein of gene YER126C is part of the 66S pre-ribosomal particles and contributes to the processing of the 27S pre-rRNA. The over-expression of this gene leads to a decrease in the vegetative growth of the yeast (Horsey et al. (2004)), which has consequences for the G1 phase of the cell cycle where the cell grows. The gene YFL021W encodes a transcription factor that activates genes involved in nitrogen catabolite repression. The gene YFL022C encodes the alpha subunit of the cytoplasmic phenylalanyl-tRNA synthetase. The over-expression of this gene is known to lead to a delay or an arrest of the G2 or M phase of the cell cycle (Niu et al. (2008)).

In summary, for both pairs of transcription factors all putative target genes predicted by the *LFC* and the *HMM* are included in the predictions of the *SHMM*(2). In comparison to the *HMM*, the specific modeling of gene pairs in head-head orientation by the *SHMM*(2) has led to the prediction of two additional cell cycle regulated target genes. The greatest number of putative common target genes that are involved in the regulation of the yeast cell cycle has been identified by the *SHMM*(2). Subsequently, the three methods are comprehensively evaluated on the ABI3 promoter array ChIP-chip data set.

7.4 Identification of Arabidopsis ABI3 Target Genes

The transcription factor ABI3 is one of the fundamental regulators of seed development involved in the control of chlorophyll degradation, storage product accumulation, and desiccation tolerance (Suzuki et al. (2003); Mönke et al. (2004); Vicente-Carbajosa and Carbonero (2005); To et al. (2006)). The inhouse promoter array ChIP-chip data set of

7. Analysis of Promoter Array ChIP-chip Data

ABI3 allows to study the target gene prediction behavior of the *LFC*, the *HMM*, and the *SHMM*(2) coupeled with validation based on publicly available expression data from Genevestigator (Zimmermann et al. (2004); Hruz et al. (2008)), and transient assays (Reidt et al. (2000)), which have been performed in wet-lab experiments during the project Arabido-Seed (2004-2009) to test whether a promoter of a putative target gene is regulated by ABI3.

7.4.1 Systematic analysis of differences between HMM and SHMM

The *HMM* approach allows to analyze ChIP-chip data in the context of chromosomal locations of genes. The application of the *SHMM*(2) extends this analysis by specifically modeling genes in head-head orientation. Here, it is investigated how the trained *SHMM*(2) behaves in comparison to the trained *HMM*. Motivated through Fig. 7.1, the global distance threshold b for the specification of the transition classes of the *SHMM*(2) is set to 9 kb, because in greater chromosomal distance the correlations of log-ratios of head-head gene pairs do not significantly differ from other gene pairs. In addition to this, each *SHMM*(2) with a scaling factor f_2 in the interval 1.1 to 10 in steps of 0.1 is studied. The State-Posterior algorithm developed in Sec. 3.4.1 is used to compare the resulting most probable state sequence of the *HMM* against the corresponding most probable state sequence of each *SHMM*(2). Here, the scaling factor f_2 allows to directly influence the prediction behavior for head-head gene pairs. That is, the greater f_2 the more likely it is that both genes of such head-head pairs are either predicted as '++' or as '−−', and the closer f_2 is set to one the more similar the prediction behavior of the *SHMM*(2) gets to that of the *HMM*. The results are summarized in Fig. 7.4. As expected, the number of head-head gene pairs for which both genes of such a pair have identical predictions increases for increasing scaling factor f_2. Consequently, a decrease of the number of head-head gene pairs for which both genes of such a pair have different predictions is observed. Obviously, each change of a head-head gene pair leads either to a change of the upstream, downstream, or both of these gene pairs. Here, the number of non-head-head gene pairs that are predicted as '++' decreases only slightly for the *SHMM*(2) with increasing scaling factor f_2 in comparison to the *HMM*. Substantially more decrease is observed for the number of non-head-head gene pairs that are predicted as '−−' for increasing scaling factor f_2. Consequently, the number of non-head-head gene pairs for which both genes of such a pair have different predictions increases with increasing scaling factor f_2. In summary, this study points out that the prediction results of the *SHMM*(2) can differ from that of the *HMM* in dependence of the value of the scaling factor f_2. Thus, the *SHMM*(2) can be seen as the more general approach.

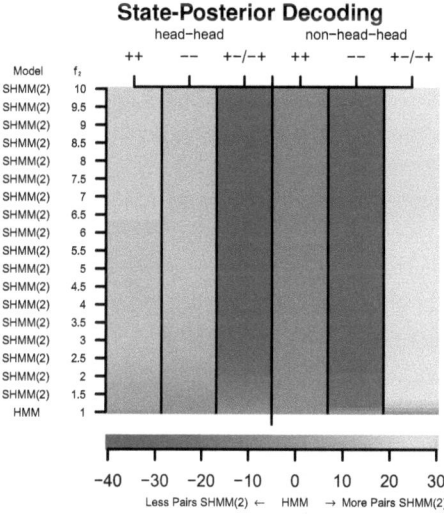

Figure 7.4: Differences of predictions of the *SHMM*(2) with scaling factor f_2 in the interval from 1.1 to 10 in steps of 0.1 in relation to the *HMM*. The *HMM* is encoded by the orange shade with value zero. The considered prediction categories of gene pairs are '++', '−−', '+−', and '−+' based on the underlying two-state architecture shown in Fig. 7.2. For $f_2 > 4$ nearly no changes in the prediction behavior are observed.

7.4.2 Comparison of ABI3 target gene predictions of LFC, HMM, and SHMM

Here, the focus is on the comparison of putative ABI3 target genes predicted by the *LFC*, the *HMM*, and the *SHMM*(2). The *LFC* method only considers the measured log-ratios of genes for the prediction of putative target genes. In contrast to this, the *HMM* also considers the chromosomal locations of genes, and the *SHMM*(2) extends this by considering gene pair orientations. That is, the goal is to find out how these three methods behave on the ABI3 data set. For that reason, the threshold for the maximal number of candidates in a top list is set to 200, because the mean log-ratio of 1.06 at this level is already relatively small, and beyond, at a top list of 300 no new putative

ABI3 target genes have been predicted by the three methods. Moreover, the trained $SHMM(2)$ with scaling factor $f_2 = 4$ is used in all further analyses, because this model is quite different from the standard HMM (Fig. 7.4), and additionally, the comparison of this model to the $SHMM(2)$ with higher scaling factors $f_2 = 6$ or $f_2 = 10$ yielded identical target genes. For each method, the top 50, 100, 150, and 200 candidates of each of the five experiments are used to determine putative ABI3 target genes. In addition to this, Venn diagrams are used to directly compare the candidate genes of these four top lists for all three methods. The results are shown in Fig. 7.5a. Here, the $SHMM(2)$ predicted the greatest number of putative ABI3 target genes, whereas the LFC method identified the smallest number. Comparing the Venn diagrams of the top 100 list to the top 200 list, all candidates that are predicted by the LFC method are also completely identified by the HMM and the $SHMM(2)$. In addition to this, the candidates additionally predicted by the HMM in the transition of the top 150 list to the top 200 list have been completely identified by the $SHMM(2)$.

Next, it is investigated whether the putative ABI3 target genes that have only been predicted by the $SHMM(2)$ at the level of the top 200 candidates are the consequence of the specific modeling of head-head orientations. For that purpose, also a $SHMM(2)$ that specifically models tail-tail orientations is trained using the identical initial settings as for the normal $SHMM(2)$. Fig. 7.5b shows that the $SHMM(2)$ that specifically models tail-tail orientations has a prediction behavior that is nearly identical to that of the standard HMM with perfect agreement at the level of the top 50 and 150 candidates, and one additional putative target gene at the level of the top 200 candidates. This coincides with the observation shown in Fig. 7.1 that the measured log-ratios of gene pairs in tail-tail orientation tend to be uncorrelated. Due to that, the specific modeling of tail-tail orientations has nearly no effect on the prediction of putative ABI3 target genes. Fig. 7.5c shows that the prediction results of the $SHMM(2)$ that specifically models tail-tail orientations are completely included in the set of predicted putative ABI3 target genes of the $SHMM(2)$ that specifically models head-head orientations. This indicates that the gain of additional putative ABI3 target genes is based on the specific modeling of head-head orientations. Taking together, the $SHMM(2)$ approach that models head-head orientations tends to be more general in the prediction of putative ABI3 target genes than the LFC, the HMM, and the $SHMM(2)$ modeling tail-tail orientations.

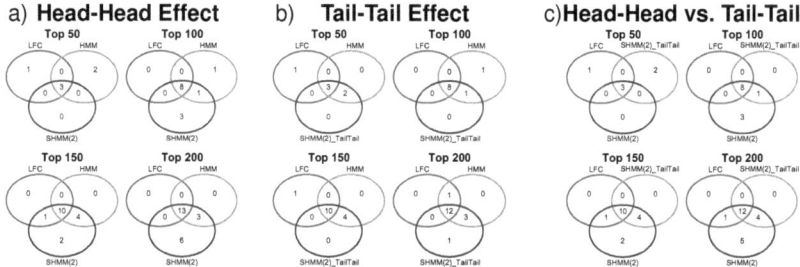

Figure 7.5: Venn diagrams for comparing the number of putative ABI3 target genes predicted by the *LFC*, the *HMM*, the $SHMM(2)$ with scaling factor $f_2 = 4.0$ that specifically models head-head orientations of genes, and the corresponding $SHMM(2)$_TailTail with $f_2 = 4.0$ that specifically models the tail-tail orientation of genes for validating the $SHMM(2)$. **a)** Comparison of the number of putative ABI3 target genes predicted by the *LFC*, the *HMM*, and the $SHMM(2)$. The $SHMM(2)$ is the most general model that predicted the greatest number of putative target genes including all genes found by the *LFC* and the *HMM* at the level of the top 150 and top 200 candidates. **b)** Comparison of putative ABI3 target genes predicted by the *LFC*, the *HMM*, and the $SHMM(2)$_TailTail. The $SHMM(2)$_TailTail does predictions nearly identical to the *HMM* with perfect agreement at the level of the top 50 and top 150 candidates. The total number of predicted putative ABI3 target genes is less than in Fig. 7.5a. **c)** Comparison of putative ABI3 target genes predicted by the *LFC*, the $SHMM(2)$_TailTail, and the $SHMM(2)$. The $SHMM(2)$ is the most general model that predicted the greatest number of putative ABI3 target genes including all genes predicted by the *LFC* and the $SHMM(2)$_TailTail. This emphasizes that the gain of additional putative ABI3 target genes is based on the specific modeling of head-head orientations by the $SHMM(2)$.

7.4.3 Biological validation of putative ABI3 target genes

Here it is investigated how putative target genes might be regulated by ABI3. For that purpose, Genevestigator (Zimmermann et al. (2004); Hruz et al. (2008)) is used as independent source of *A. thaliana* gene expression data to analyze the predicted putative target genes. In Genevestigator, ABI3 is mainly expressed within the categories inflorescence, silique, and seed. Based on that, the gene expression level of each putative target gene is quantified by dividing the sum of its gene expression levels within these three categories by the sum of gene expression levels in all categories. This provides a quantitative measure subsequently referred to as Genevestigator score for analyzing how a putative ABI3 target gene follows the expression profile of ABI3. Additionally, transient assays have been performed in wet-lab experiments at the IPK Gatersleben to test whether the promoters of putative ABI3 target genes in fusion with the glucuronidase (GUS) reporter gene react on ABI3. The results are shown in Tab. 7.2 using anonymized gene identifiers because the work on a biological manuscript is still in progress. Based on the Genevestigator score, 16 of 22 putative target genes show significantly high scores at the level of the 95%-quantile 0.15 computed using the Genevestigator scores of 1,000 randomly selected genes. The promoters of these 16 genes have been tested in transient assays, and 15 of these promoters can activate the GUS expression through ABI3. The promoter of gene T21 shows nearly a two-fold repression of the GUS expression, which is not reflected by its Genevestigator score. Interestingly, the genes T21 and T22 are in head-head orientation to each other, and thus they have the potential to share a common promoter region. Based on the results of the transient assays, the first gene might be repressed while the second is activated. Hence, it seems that activation and repression signals can be transmitted by ABI3 to these two target genes in head-head orientation via a common promoter region. Additionally, 3 of these 15 target genes activated by ABI3 and the one with a nearly two-fold repression have only been predicted by the $SHMM(2)$. In contrast to these 16 target genes, the 6 remaining putative target genes do not significantly differ in their Genevestigator scores at the level of the 5%-95%-quantile range [0.02, 0.15] based on the distribution of the Genevestigator scores for the 1,000 randomly selected genes. Interestingly, 5 of these 6 putative target genes are in head-head orientation to one of the previous target genes activated by ABI3. Next, the question if these 6 putative ABI3 target genes are also under control of ABI3 is addressed. To analyze this, the promoters of 4 of these 6 putative target genes have been tested in transient assays. The promoters of the genes T2 and T11 show a low activation of the GUS expression, the promoter of gene T13 shows a two-fold repression of the GUS expression, and the promoter of gene T9 does not seem to react on ABI3. In addition to this, gene T13 is in head-head orientation with gene T23 that is not represented by its own promoter fragment on the promoter arrays. The Genevestigator score of T23 is significantly higher than those of the 1,000 random genes at the level of the 95%-quantile, and the promoter of this gene shows activation of the GUS expression in a transient assay. Hence, this gene pair seems to behave like the gene pair T21 and T22.

In summary, independent gene expression profiles from Genevestigator give first hints which genes might be activated by ABI3. Additionally, transient assays help to validate these results if the underlying test system is capable of simulating the natural situation in seeds. In total, 20% of the target genes with high Genevestigator scores and activation by ABI3 could be predicted only through the application of the $SHMM(2)$ and

7. Analysis of Promoter Array ChIP-chip Data

ID	LFC	HMM	SHMM	GV	TA	ID	LFC	HMM	SHMM	GV	TA
T1	1	1	1	0.94	5	T20	1	1	1	0.57	8
T2	1	1	1	0.11	2.5	T22	1	1	1	0.81	30
T3	1	1	1	0.86	12	T11	0	1	1	0.09	2
T6	1	1	1	0.72	15	T15	0	1	1	0.10	-
T7	1	1	1	0.90	7	T18	0	1	1	0.98	27
T12	1	1	1	0.74	24	T4	0	0	1	0.03	-
T13	1	1	1	0.09	0.4	T5	0	0	1	0.39	3
T14	1	1	1	0.93	8	T8	0	0	1	0.46	12
T16	1	1	1	0.95	27	T9	0	0	1	0.07	1
T17	1	1	1	0.98	27	T10	0	0	1	0.95	6
T19	1	1	1	0.98	27	T21	0	0	1	0.20	0.6

Table 7.2: Overview of ABI3 target genes predicted by the *LFC*, the *HMM*, and the *SHMM*(2) with scaling factor $f_2 = 4.0$ at the level of the top 200 candidates in Fig. 7.5a. The ID column contains anonymized target gene identifiers because a biological manuscript discussing individual target genes is still in preparation. The numbers '1' and '0' in the method columns *LFC*, *HMM*, and *SHMM*(2) specify whether a gene is predicted ('1') or missed ('0'). GV (Genevestigator score) quantifies the gene expression of a target gene within the categories inflorescence, silique, and seed in relation to the gene expression levels in all categories. TA (Transient assay) contains the measured fold-change of the GUS gene expression for a target gene promoter under ABI3 expression in relation to this target gene promoter lacking the expression of ABI3.

would have been missed using the *LFC* or the *HMM*. Thus, the integration of additional information about gene pair orientations into the analysis of ABI3 promoter array ChIP-chip data by the *SHMM*(2) has led to improved predictions of target genes. Also with respect to the results obtained for the promoter array ChIP-chip of the yeast, the *SHMM*(2) is a useful tool for the detection of target genes that could be considered for the analysis of other data sets.

7.5 Further reading

Parts of this study were presented at the German Conference on Bioinformatics (Seifert et al. (2008)) and have later been published (Seifert et al. (2009b)). An additional study analyzing ABI3 target genes has recently been published (Mönke et al. (2012)) utilizing the *SHMM* as the basic model for target gene predictions. Additionally, the *SHMM* has been utilized to identify target genes of the transcription factor LEC1 (Junker et al. (2012)). Moreover, the basic two-state *HMM* has been included into a comparison study for the analysis of DNA methylation profiles measured on tiling arrays (Seifert et al. (2012b)). Java-based implementations of the *HMM* and the *SHMM* are publicly available from http://www.jstacs.de/index.php/SHMM.

8 Analysis of Arabidopsis Array-CGH Data

The method of array-based comparative genomic hybridization (Array-CGH) has been widely applied to detect sequence polymorphisms like deletions or amplifications of DNA segments between two genomes (Pinkel and Albertson (2005)). Currently, most of the Array-CGH studies have their focus in cancer research (Beroukhim et al. (2010)). Regarding the field of plant research, studies like those by Borevitz et al. (2003), Martienssen et al. (2005), or Fan et al. (2007) have been done for the model plant *Arabidopsis thaliana*. In this chapter, an Array-CGH data set of *A. thaliana* comparing the genomes of the two accessions C24 and Columbia (Col) is analyzed. This data set is part of the experiments done by Banaei (2009) at the IPK Gatersleben utilizing a whole-genome tiling array that represents the reference genome of Col that has been sequenced by The Arabidopsis Initiative (2000). The whole-genome tiling array is a high-density DNA microarray that contains thousands of genomic regions (tiles) of Col for investigating their behavior in other accessions like C24. This tiling array enables the identification of sequence polymorphisms in C24 relative to the genome sequence of Col. With respect to the reference genome of Col, detectable sequence polymorphisms comprise genomic regions that are deleted or highly polymorphic in C24 and genomic regions that are amplified in C24. Generally, an Array-CGH data set created on a tiling array does not allow to distinguish between deleted or highly polymorphic genomic regions. Both types of these sequence polymorphisms are associated with reduced signal intensities. In the context of C24, deleted genomic regions are less frequently present in the genome of C24 in comparison to their occurrence in Col. For highly polymorphic genomic regions in C24, the tiles representing these genomic regions in Col are affected in C24 by single nucleotide polymorphisms or small insertions and deletions. Thus, deleted or highly polymorphic regions in C24 are expected to have lower hybridization signals than in Col. In contrast to this, genomic regions that are amplified in C24 are more frequently present in the genome of C24 in comparison to Col leading to higher hybridization signals for C24 than for Col. To provide an overview, the measurements of the Array-CGH data set comparing the genomes of C24 and Col and the dependencies of these measurements in the context of the underlying chromosomal locations are shown in Fig. 8.1. Based on the histogram of measured log-ratios (Fig. 8.1 left), the majority of the genomic regions of C24 and Col is expected to be unchanged. In addition to this, a large proportion of tiles tends to be deleted or highly polymorphic in C24, whereas only a small fraction of tiles is expected to represent genomic regions that are amplified in C24. The general goal is to identify these sequence polymorphisms in the context of their chromosomal locations. Higher-order dependencies between measured log-ratios of adjacent tiles in close chromosomal proximity are clearly observed for the Array-CGH data set (Fig. 8.1 right). The observation of these dependencies motivates the usage of the higher-order *HMM* ($HHMM(L)$) and the parsimonious higher-order *HMM* ($PHHMM(L)$). Both models integrate these dependencies into the prediction of sequence polymorphisms of the Array-CGH data set. To investigate whether the modeling of higher-order dependencies improves the prediction of sequence polymorphisms, the $HHMM(L)$ and the $PHHMM(L)$ are compared to the standard first-order *HMM* that only considers dependencies between two directly adjacent tiles on a chromosome. Besides this, further comparisons are made against frequently used methods for the analysis of Array-CGH data summarized in Tab. 6.1 provided by the ADaCGH web-server (Diaz-Uriarte and Rueda (2007)). All these models are compared based on deleted or highly polymorphic genomic regions that have been identified in the genome of C24 with respect to Col using two independent resequencing technologies. In the frame of a cooperation at the IPK Gatersleben, deleted or highly polymorphic genomic regions determined by SOLiD resequencing (Applied Biosystems (2009)) of C24 are considered for the validation of the models. Additionally, publicly available deleted or highly polymorphic regions in C24 identified in Affymetrix resequencing data for C24 by Clark et al. (2007) are considered too. Thus, these two validation data sets provide the opportunity to investigate which model is appropriate for the analysis of the Array-CGH data set. Moreover, it is also analyzed what is functionally behind the genomic regions in which the genomes of C24 and Col differ.

Goals of this Chapter

1. More details to the Array-CGH data set for comparing the genomes of the accessions C24 and Col are given.

8. Analysis of Arabidopsis Array-CGH Data

2. *HMM*-based approaches for the analysis of the Array-CGH data set are developed and related methods are briefly summarized.

3. The potential of the $HHMM(L)$ for modeling the higher-order dependencies of log-ratios in the Array-CGH data set is investigated.

4. The SOLiD and Affymetrix resequencing data sets for the validation of the predictions on the Array-CGH data set are introduced.

5. The performances of the *HMM*, the $HHMM(L)$, and the $PHHMM(L)$ on the Array-CGH data set are evaluated and compared to related methods.

6. Selected tree structures that underlie the $PHHMM(L)$ are investigated and the predictions are analyzed in the context of the genome annotation.

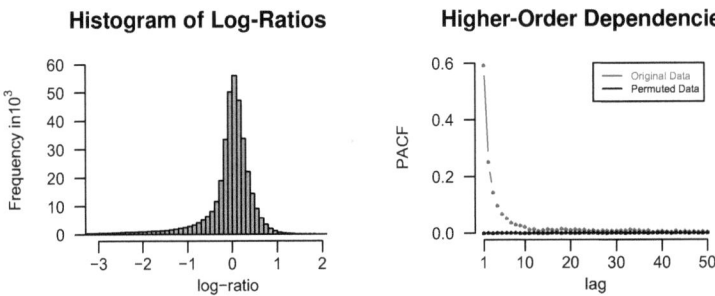

Figure 8.1: Overview of the Array-CGH data set by Banaei (2009) comparing the genomes of C24 and Col. Left: Histogram of log-ratios for the 364,339 measured tiles (genomic regions). The log-ratio of a tile represents the ratio between the DNA fragments of C24 compared to the DNA fragments of Col hybridized to this tile on the DNA microarray. Most parts of the genomes of C24 and Col tend to be unchanged with log-ratios of about zero. Deleted or highly polymorphic genomic regions in C24 have log-ratios much less than zero, and amplified regions in C24 tend to have log-ratios much greater than zero. Right: Higher-order dependencies between log-ratios measured for adjacent tiles in close chromosomal proximity characterized by partial autocorrelations (Gottman (1981)). The weighted mean partial autocorrelation function (PACF) of the log-ratios of the five chromosomes in the Array-CGH data set is shown for an increasing lag of tiles. The PACF of a chromosome is weighted by its proportion of log-ratios in relation to the total number of log-ratios in the Array-CGH data set. Positive higher-order partial autocorrelations are clearly present in the data set. These dependencies are lost if the log-ratios of a chromosome are randomly permuted.

8. Analysis of Arabidopsis Array-CGH Data

8.1 Arabidopsis Array-CGH Data Set

The Array-CGH data set has been provided by Banaei (2009) within a cooperation at the IPK Gatersleben. The comparison of the genomes of the *A. thaliana* accessions C24 and Col has been done based on DNA extracted from leaf tissue using the Array-CGH technology reviewed by Pinkel and Albertson (2005). The obtained DNA of each accession has been sheared into smaller DNA segments of lengths between about 300 bp up to 900 bp. The resulting DNA segments have been labeled by a specific fluorescent dye for each accession. The obtained labeled single-stranded DNA segments of both accessions have then been hybridized simultaneously in a competitive style to a NimbleGen tiling array that represents the reference genome of accession Col. The reference genome of Col has been sequenced by The Arabidopsis Initiative (2000). The size of this genome comprises about 119 Mb (TAIR8). The tiling array represents the genome of Col by 364,339 single-stranded DNA fragments (tiles) of sizes about 60 bp. The tiles are distributed nearly equidistantly over the five chromosomes with a mean distance of 350 bp for two adjacent tiles on a chromosome. The measurement obtained for each tile t of chromosome k on the tiling array is the normalized log-ratio $o_t(k) := \log_2(I_t^k(\text{C24})/I_t^k(\text{Col}))$ of the fluorescent intensity $I_t^k(\text{C24})$ measured for C24 in relation to the fluorescent intensity $I_t^k(\text{Col})$ measured for Col. All T_k log-ratios of a chromosome k are represented in increasing order of their chromosomal locations by the emission sequence $\vec{o}(k) = (o_1(k), \ldots, o_{T_k}(k))$. In total, five emission sequences that represent the five chromosomes of *A. thaliana* are obtained from the tiling array. An overview of the log-ratios in the Array-CGH data set is given by the histogram in Fig. 8.1. The asymmetry of this histogram might be caused due to the design of the tiling array that only represents DNA segments of the reference genome of Col meaning that DNA segments that only exist in C24 but not in Col cannot be quantified by this approach.

8.2 Methods for Array-CGH Data Analysis

8.2.1 Hidden Markov Model approaches

Models: A basic three-state architecture shown in Fig. 8.2 with states $S := \{-, =, +\}$ and state-specific Gaussian emission densities is used to identify sequence polymorphisms in C24 in comparison to Col. With respect to the log-ratios shown in Fig. 8.1,

8. Analysis of Arabidopsis Array-CGH Data

non-polymorphic regions with log-ratios about zero are modeled by the state '='. The state '−' models deleted or highly polymorphic genomic regions in C24 represented by log-ratios much less than zero, and amplifications in C24 with log-ratios much greater than zero are modeled by the state '+'. The three-state architecture in Fig. 8.2 represents the basis of the HHMM(L) and the PHHMM(L). Generally, the usage of these models is motivated through the observation of higher-order dependencies of log-ratios on a chromosome. As indicated by highly positive partial autocorrelations in Fig. 8.1, these dependencies are clearly present in the Array-CGH data set. The HHMM(L) and the PHHMM(L) are applied to this data set by considering each order L in the range of one up to five. Like specified in Tab. 3.1, the HMM is as a special case of the HHMM(L) of order $L = 1$. Models with additional transition classes like the HHMM(L, C) or the PHHMM(L, C) are not considered, because the tiles measured in the Array-CGH data set are distributed nearly equidistantly along the chromosomes. Thus, the most general model applied here is the PHHMM(L) that includes the HMM and HHMM(L) as special cases.

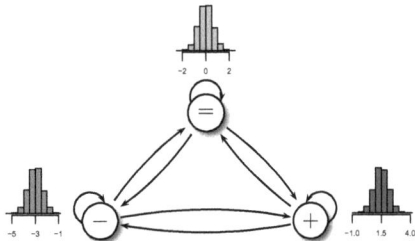

Figure 8.2: The basic three-state architecture of the HMM, the HHMM(L), and the PHHMM(L) used for the analysis of Array-CGH data. The states $S := \{-, =, +\}$ are represented by labeled circles and corresponding state-specific Gaussian emission densities. Arrows represent the possible state transitions.

Prior: The prior defined in (3.29) provides the basics for each model to distinguish between polymorphic and non-polymorphic regions with respect to the given log-ratios in the Array-CGH data set. A histogram of log-ratios like shown in Fig. 8.1 helps to characterize the states of each model by proper prior parameters. Based on this, the means of the Gaussian emission densities of the emission prior defined in (3.32) are set to $\eta_- = -3$, $\eta_= = 0$, and $\eta_+ = 1.5$. Motivated through the observation that most log-ratios are distributed around zero and that the distribution of the log-ratios has a larger left tail than a right one, the scale parameters $\epsilon_- = \epsilon_= = 1$ and $\epsilon_+ = 5,000$ are used to provide more flexibility for the training of the means of the states '−' and '=' than for the mean of the state '+'. The shape parameter r_i and the scale parameter α_i of the emission prior of each state $i \in S$ are set in dependence of the number of log-ratios $T = 364,339$ in the Array-CGH data set and their standard deviation $s = 0.668$ to $r_i = T/2$ and $\alpha_i = T \cdot s^2/2$. The parameter ϑ_i of the prior of the initial state distribution defined in (3.30) is set to $\vartheta_i = 3^L$ for each state $i \in S$. For the transition prior given in (3.31), the parameter $\vartheta_{ij}(1) = 3^{L-t}$ is used for each state context $i \in S^t$ of length $1 \leq t \leq L$ and each next state $j \in S$. Finally, the parameter φ of the tree structure prior (4.7) must be specified for the *PHHMM(L)* to enable the parsimonious representation of the transition parameters. The values of φ are generally depending on the size of the data set. All initial models have completely fused trees for $\log(\varphi) = -30,000$. The range of $\log(\varphi)$ from $-23,000$ up to 0 has been considered to obtain parsimonious trees. This range is considered in steps of $1,000$ for $-23,000$ to $-1,000$, in steps of 100 for $-1,000$ to -100, and finally from -100 up to 0 in steps of 10. Complete trees that underlie the *HMM* and the *HHMM(L)* are obtained for each initial model by $\log(\varphi) = 10$ at latest.

Initialization: Each initial model in Fig. 8.2 should be able to distinguish between non-polymorphic regions modeled by the state '=' and polymorphic regions represented by the states '−' and '+'. For that reason, the initial means of the Gaussian emission densities are set to $\mu_- = -3$, $\mu_= = 0$, and $\mu_+ = 1.5$ with respect to the initial overview of the Array-CGH data set shown in Fig. 8.1. The corresponding initial standard deviation σ_i of each state $i \in S$ is set to the standard deviation of the log-ratios in the Array-CGH data set given by 0.668. The parameters of the initial state distribution and the transition parameters of each model are sampled from the corresponding start and transition prior.

Training: For the training of each *PHHMM(L)* in combination with each of the previously specified values of $\log(\varphi)$ twenty different initial models are considered for each combination. Each *PHHMM(L)* of order L ranged from one up to five is trained using the Bayesian Baum-Welch algorithm developed in Sec. 4.4. This training algorithm includes the biological prior knowledge specified by the prior. The training of each *PHHMM(L)* is stopped if the improvement of the log-posterior of two successive iteration steps is less than 10^{-9}. The *HMM* and the *HHMM(L)* are obtained as special cases of the *PHHMM(L)* with complete tree structures.

Detection of deleted or highly polymorphic genomic regions: Putatively deleted or highly polymorphic genomic regions in C24 are expected to have negative log-ratios in the Array-CGH data set (Fig. 8.1). To quantify the potential that a tile represents

8. Analysis of Arabidopsis Array-CGH Data

these sequence polymorphisms, a score is computed for each tile under the *HMM*, the *HHMM*(L), or the *PHHMM*(L). The score of a tile t on chromosome k is computed via the state-posterior $\gamma_t^k('-')$ given in (3.6). The state-posterior quantifies the probability of a tile to be represented by the state '−' modeling deleted or highly polymorphic regions. Based on this score, all tiles in the Array-CGH data set are ranked according to their potential to represent a deleted or highly polymorphic genomic region. The ranked tiles can be associated via their genomic locations to deleted or highly polymorphic regions known from independent validation experiments. This allows to compare the prediction behavior of different models for fixed false positive rates.

8.2.2 Related approaches for the analysis of Array-CGH data

The analysis of the Array-CGH data set shown in Fig. 8.1 by publicly available methods provides the opportunity for the comparison to the *HMM*-approaches. The standard method for the analysis of the Array-CGH data set measured on a NimbleGen tiling array is the segMNT algorithm by Roche NimbleGen, Inc. (2008). Additionally, the ADaCGH web-server by Diaz-Uriarte and Rueda (2007) provides different methods from the field of cancer research that can be tested on the Array-CGH data set. A summary of methods of the ADaCGH web-server is given in Tab. 6.1. The segMNT algorithm and all methods provided by the ADaCGH web-server are used to predict deleted or highly polymorphic genomic regions in C24. The obtained prediction results are compared to the developed *HMM*-approaches.

8.3 Arabidopsis Array-CGH Data Analysis

8.3.1 Analysis of dependencies between log-ratios

The log-ratios measured for the tiles along a chromosome in the Array-CGH data set represent a spatial sequence of data in the genomic context. The linear dependency of two log-ratios in a lag of l tiles with removed linear dependency of all log-ratios between these two log-ratios is quantified by the value of the partial autocorrelation function (PACF) at lag l. The PACF at lag l can be computed like described in the textbook of Gottman (1981) through the estimation of an autoregressive model of order l. An autoregressive model of order l estimated from a sequence of data represents all values of the PACF of this sequence up to lag l, and for greater lags the PACF is zero.

The PACF is specifically useful in time series analysis for determining the order of an autoregressive model (Gottman (1981)).

Here, it is investigated how the PACF of the Array-CGH data set can be modeled by the $HHMM(L)$ of order L ranged from zero up to five. Each trained $HHMM(L)$ has been used to sample 100 emission sequences with 10,000 log-ratios. Based on this, the PACF has been computed for the sampled emission sequences of each $HHMM(L)$. The results are shown in Fig. 8.3. As expected from theory, the mixture model of Gaussian emission densities (e.g. Bilmes (1998)) represented by the $HHMM(0)$ does not model dependencies between two log-ratios in any lags. The modeling of dependencies by the standard first-order $HHMM(1)$ clearly improves the representation of the PACF of the Array-CGH data set. This behavior is further improved by the usage of the $HHMM(2)$ for lags greater than one. The best models are the $HHMM(3)$, the $HHMM(4)$, and the $HHMM(5)$ that better model the PACF of the Array-CGH data set than the $HHMM(2)$ for lags greater than two. However, especially for lag one all these models clearly underestimate the PACF of the Array-CGH data set. One reason for this is the difference between the hybridization of DNA segments to create the log-ratios measured for the Array-CGH data set and the sampling of log-ratios from state-specific Gaussian emission densities. The DNA segments hybridized to the tiles have lengths up to 900 bp. Thus, log-ratios measured for directly adjacent tiles on a chromosome with distance about 350 bp are expected to be more similar to each other than log-ratios that are sampled from a state-specific Gaussian emission density that has to cover a larger range of log-ratios. The PACF of the Array-CGH data set could of cause be modeled perfectly by an autoregressive model, but this model class does not allow to predict deleted or highly polymorphic genomic regions as well as amplified genomic regions. However, this prediction can be done directly using HMM-based approaches.

8.3.2 SOLiD and Affymetrix resequencing data for validating the Array-CGH data set

Two large-scale resequencing data sets by SOLiD and Affymetrix provide the opportunity to validate deleted or highly polymorphic regions in C24 that have been predicted in the Array-CGH data set. SOLiD is one of the next-generation sequencing technologies (Shendure et al. (2005); Mardis (2008); Applied Biosystems (2009)) that has been used to resequence the genome of C24 (Prof. Dr. T. Altmann, IPK Gatersleben, unpublished

8. Analysis of Arabidopsis Array-CGH Data

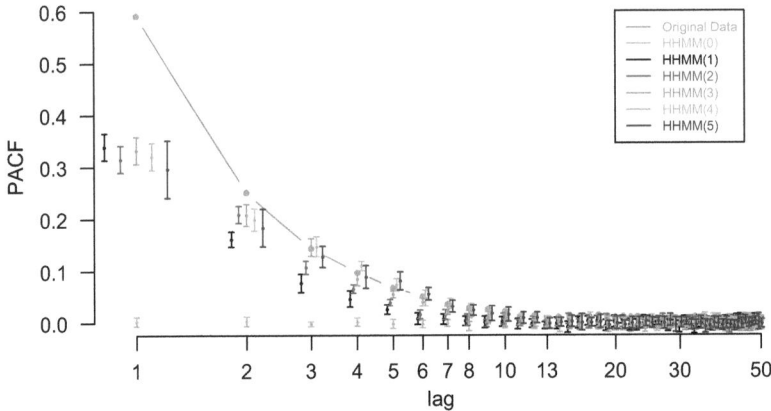

Figure 8.3: Overview of the partial autocorrelation function (PACF) obtained for the Array-CGH data set and for data sampled from the trained $HHMM(L)$. Here, the value of the PACF at lag l is defined to be a weighted mean specified by the sum over each individual PACF of a chromosome at lag l weighted by its proportions of log-ratios in relation to the total number of log-ratios in the Array-CGH data set. The PACF of the log-ratios measured for the five chromosomes in the data set is shown in orange. The PACF shown for each $HHMM(L)$ is computed based on 100 artificial chromosomes with 10,000 log-ratios sampled from this model. The $HHMM(0)$ is a mixture model of Gaussian densities that does not model dependencies between adjacent log-ratios on a chromosome. The $HHMM(1)$ is the standard first-order HMM. To ease the comparison, the lag-axis is plotted in logarithmic scale, and the PACFs computed for the $HHMM(L)$ are plotted slightly shifted for of each lag.

8. Analysis of Arabidopsis Array-CGH Data

data). Affymetrix resequencing is based on high-density oligonucleotide arrays and has been applied by Clark et al. (2007) to resequence twenty accessions of *A. thaliana* including C24. Subsequently, the results obtained by SOLiD and Affymetrix are used to create two independent validation data sets for the Array-CGH-based genome comparison of C24 to Col.

SOLiD resequencing data of C24

The genome of C24 is represented in the SOLiD data set by reads of length 35 bp. All these reads have been mapped back to the TAIR8 reference genome sequence of Col using the standard software of Applied Biosystems (2009). This software considers a genomic region of 35 bp in Col as covered by a SOLiD read of C24 if the read of C24 and the genomic region of Col do not differ in more than 3 bp. Genomic regions of Col that have not been covered by reads of C24 are strong candidates of deleted or highly polymorphic regions for C24 in comparison to Col. These candidate regions have been mapped to the 364,339 tiles of 60 bp length that represent the genome of Col in the Array-CGH data set. Each tile that has been covered to a certain percentage by such a candidate region is labeled as being putatively deleted or highly polymorphic. In the following, a stringent and a less stringent coverage are considered for the validation of the Array-CGH data. A stringent coverage of at least 75% (at least 45 of 60 bp of a tile) results in 38,567 tiles labeled as deleted or highly polymorphic and 325,772 non-labeled tiles. Subsequently, this labeled Array-CGH data set is denoted as the SOLiD 75% validation data set. For a less stringent coverage of at least 40% (at least 24 of 60 bp of a tile), 50,397 tiles have been labeled as deleted or highly polymorphic and 313,942 tiles remain non-labeled. This data set is denoted as the SOLiD 40% validation data set. As indicated in Fig. 8.4, a large proportion of tiles labeled as deleted or highly polymorphic is associated with log-ratios much less than zero in the Array-CGH data set. Thus, a large proportion of these tiles should also be predicted as deleted or polymorphic in the Array-CGH data set.

Affymetrix resequencing data of C24

The Affymetrix resequencing data of C24 created by Clark et al. (2007) has been further analyzed by Zeller et al. (2008) resulting in deleted or highly polymorphic candidate regions in C24 in comparison to the TAIR8 reference genome sequence of Col. These candidate regions have been used to identify all tiles in the Array-CGH data set that are covered by at least 45 bp (75% coverage) or by at least 24 bp (40% coverage). Under the 364,339 tiles in the Array-CGH data set, 11,025 tiles have been labeled as deleted or highly polymorphic and 353,314 tiles remain non-labeled for a coverage of 75%. Subsequently, it is referred to this data set as the Affymetrix 75% validation data. In analogy, the Affymetrix 40% validation data set is specified by 24,231 tiles that have been labeled as deleted or highly polymorphic and 340,108 non-labeled tiles at a less stringent coverage of 40%. Fig. 8.4 shows that the Affymetrix resequencing data provides the opportunity to validate deleted or highly polymorphic genomic regions of the Array-CGH data set. Most of the labeled tiles have log-ratios much less than zero as expected from Fig. 8.1 for putatively deleted or highly polymorphic regions in C24.

Notes on SOLiD and Affymetrix data sets

The reference genome of accession Col has about 119 Mb. About 16.6 Mb of this reference genome have not been covered by reads of C24 using the SOLiD resequencing technology. Considering the results obtained for the Affymetrix resequencing technology, about 7.9 Mb in C24 are deleted or highly polymorphic in comparison to Col. That means, the SOLiD technology has provided more than twice the number of putatively deleted or polymorphic DNA bases than Affymetrix. On the one hand, the high number of deleted or polymorphic DNA bases identified by SOLiD could be biased through the short read length of 35 bp, sequencing errors or highly polymorphic reads that could not be mapped back to the reference genome of Col. On the other hand, the number of deleted or polymorphic DNA bases provided by Zeller et al. (2008) for the Affymetrix resequencing data set of C24 might be to small, because their method performed better for coding sequences that have a higher GC content and sequence complexity than non-coding sequences. The differences of both resequencing technologies can also be clearly seen in Fig. 8.4 for the mapping of putatively deleted or highly polymorphic regions to the corresponding tiles in the Array-CGH data set. Here, a certain proportion of putatively deleted or highly polymorphic regions of the SOLiD validation data is associated with log-ratios about zero in the Array-CGH data set. Thus, based on the Array-CGH data set, one would expect that the underlying genomic regions are not deleted or highly polymorphic between C24 or Col. Reasons for these differences can be limitations in the mapping of reads due to the short read length, sequencing errors or highly polymorphic reads occurring in SOLiD data sets like described by Ondov et al. (2008). Thus, one cannot expect to predict many of these putatively deleted or highly polymorphic regions with log-ratios about zero in the Array-CGH data set. Anyhow, both the SOLiD and the Affymetrix resequencing data sets provide the opportunity to independently validate predictions of deleted or highly polymorphic genomic regions in the Array-CGH data set.

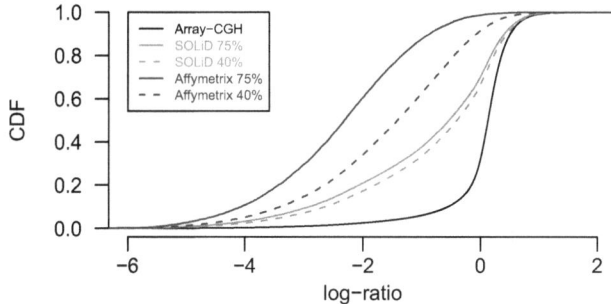

Figure 8.4: Overview of the measured log-ratios of tiles in the Array-CGH data set that have been identified as deleted or highly polymorphic by SOLiD or Affymetrix resequencing. A tile that is covered to at least 75% or 40% by a deleted or highly polymorphic region of a resequencing experiment is labeled as deleted or highly polymorphic in the Array-CGH data set. The cumulative distribution of all log-ratios of the Array-CGH data set is shown in black, that for log-ratios of tiles that have been labeled as deleted or highly polymorphic by SOLiD are shown in green, and that for log-ratios of tiles labeled as deleted or highly polymorphic by Affymetrix are shown in blue. Putatively deleted or highly polymorphic regions given by SOLiD or Affymetrix are clearly associated with negative log-ratios measured in Array-CGH. The resulting cumulative distributions for the two independent resequencing technologies SOLiD and Affymetrix are clearly different.

8.3.3 Performance of HHMMs on the Array-CGH data set

Putatively deleted or highly polymorphic genomic regions of C24 in the Array-CGH data set are modeled by the state '−' of the three-state architecture of the *HMM* and the *HHMM(L)* shown in Fig. 8.2. The potential that a tile in the Array-CGH data set represents such a sequence polymorphism is quantified by the probability that this tile is modeled by state '−'. Based on that, all tiles in the Array-CGH data set have been ranked by decreasing values of the corresponding probabilities. The tiles in the resulting ranking list that represent putatively deleted or highly polymorphic regions in C24 are known from the SOLiD and the Affymetrix resequencing data. This allows to compare the *HMM* and the *HHMM(L)* based on the true positive rate (TPR)

8. Analysis of Arabidopsis Array-CGH Data

of predicted deleted or highly polymorphic regions reached for a fixed false positive rate (FPR). The mean TPRs obtained for the twenty different initializations of the *HMM* and the *HHMM*(L) for a fixed FPR of 1% are shown in Fig. 8.5 separately for the SOLiD and the Affymetrix resequencing validation data. For the stringent SOLiD 75% validation data, the standard first-order *HMM* performs better than the higher-order *HHMM*(L) that shows a decreasing mean TPR for increasing order L. This behavior has changed completely for the less stringent SOLiD 40% validation data. Tiles that are not labeled as deleted or highly polymorphic in the stringent SOLiD 75% validation data set are indeed covered by a deleted or highly polymorphic region in the less stringent SOLiD 40% validation data set. These deleted or highly polymorphic regions are also present in the measured log-ratios of the corresponding tiles in the Array-CGH data set. The higher-order *HHMM*(L) is better able to recognize these effects on the log-ratios than the standard first-order *HMM*. In more detail, the best model for the SOLiD 40% validation data set is the fourth-order *HHMM*(4), and the fifth-order *HHMM*(5) is on the level of the third-order *HHMM*(3) and the second-order *HHMM*(2). All these higher-order *HHMM*(L) perform clearly better than the standard first-order *HMM*. Considering the Affymetrix resequencing data mapped to the Array-CGH data set, for both, the Affymetrix 75% and the Affymetrix 40% validation data sets, the higher-order *HHMM*(L) performs clearly better than the standard first-order *HMM*. For the Affymetrix 75% validation data set, the fourth-order *HHMM*(4) is the best model, and the fifth-order *HHMM*(5) is clearly below the second-order *HHMM*(2). The best model on the less stringent Affymetrix 40% validation data set is the fifth-order *HHMM*(5) that is slightly better than the fourth-order *HHMM*(4). Generally, the usage of the *HHMM*(L) shows a clear improvement of the prediction of deleted or highly polymorphic regions in C24 in comparison to the standard first-order *HMM* except for the stringent SOLiD 75% validation data. Subsequently, it is investigated whether this improvement can be further increased by using the *PHHMM*(L).

8.3.4 Performance of PHHMMs on the Array-CGH data set

The *PHHMM*(L) provides the opportunity to reduce the model complexity of the *HHMM*(L) and this might effect the prediction of deleted or highly polymorphic regions in C24 for the Array-CGH data set. The complexity of a model is given by the number of leaves in its corresponding tree that represents the transition parameters for state contexts $i \in S^L$ of length L. The model complexity of each *PHHMM*(L) of initial order L is controlled via the value φ of the tree structure prior. For the prediction of

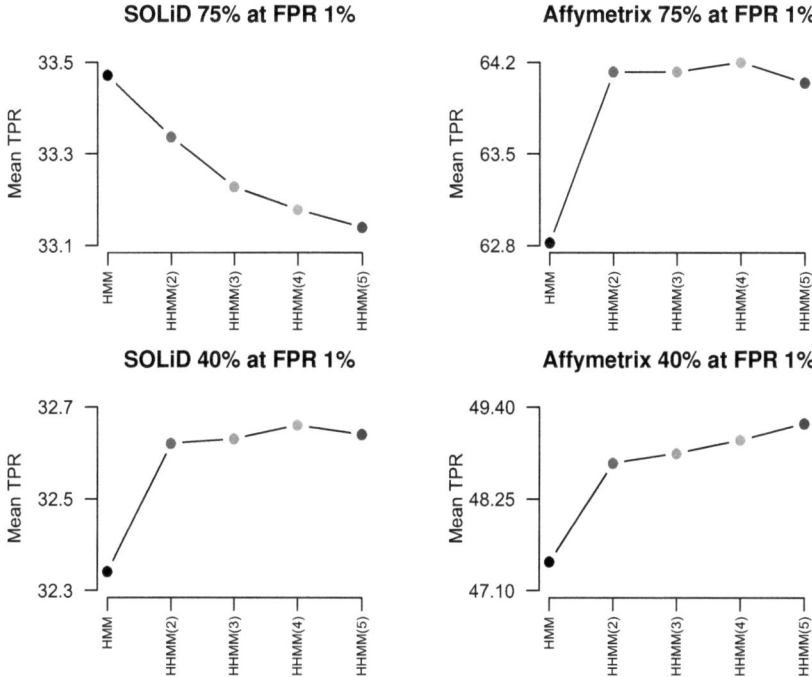

Figure 8.5: Overview of the mean TPRs obtained at a fixed FPR of 1% for the twenty different initializations of the *HMM* and the *HHMM*(L) of order $L = 2$ up to $L = 5$ on the Array-CGH data set under consideration of the SOLiD and the Affymetrix resequencing validation data. The top graphics represent the results obtained for the stringent coverage of 75% and the bottom graphics show the results obtained for the less stringent coverage of 40%. The obtained standard deviations of the TPRs were smaller than the size of the shown points.

8. Analysis of Arabidopsis Array-CGH Data

deleted or highly polymorphic regions, the ranking list of the tiles has been created for each $PHHMM(L)$ as described for the $HHMM(L)$ in the previous section. To investigate how the $PHHMM(L)$ behaves in comparison to the results of the HMM and the $HHMM(L)$ shown in Fig. 8.5, the mean TPRs have been determined separately at the same level of 1% FPR for the SOLiD and the Affymetrix validation data. All $PHHMM(L)$ with model complexity of 1 had have mean TPRs much less than those obtained for the HMM in Fig. 8.5. Thus, these models are not further considered. The results of all other $PHHMM(L)$ are shown in Fig. 8.6. For both, the SOLiD and the Affymetrix validation data, the mean TPR at a fixed FPR of 1% is improved in almost all cases by the application of a $PHHMM(L)$ in comparison to the results obtained for the $HHMM(L)$ shown in Fig. 8.5. The best models for the SOLiD 75% validation data have a model complexity of two leaves, which is one leaf less than represented by the complete tree in Fig. 4.1 that is underlying the first-order HMM. Thus, for this validation data set highly parsimonious models like the $PHHMM(1)$ with a model complexity of two leaves perform best. However, like motivated in the previous section, the SOLiD 75% validation data set tends to be too stringent. This is clearly indicated by the results obtained for the less stringent SOLiD 40% validation data, for which the higher-order $PHHMM(L)$ performs better than the first-order $PHHMM(1)$. The best models for the SOLiD 40% validation data set are among the $PHHMM(4)$ and the $PHHMM(5)$ with a model complexity of more than 27 leaves. Considering the Affymetrix validation data, the mean TPR at a fixed FPR of 1% is clearly improved by the $PHHMM(L)$ in comparison to the corresponding $HHMM(L)$ for the stringent Affymetrix 75% validation data. The best models of the $PHHMM(2)$ up to the $PHHMM(5)$ have a mean model complexity in the range of about 3 up to 9 leaves. In this range, the best mean TPRs are obtained for the $PHHMM(4)$ and the $PHHMM(5)$. For the less stringent Affymetrix 40% validation data set, the best models among the $PHHMM(2)$, the $PHHMM(3)$, and the $PHHMM(5)$ obtain mean TPRs comparable to that obtained for the corresponding $HHMM(L)$ by having a much lower model complexity. Only the $PHHMM(4)$ represents models that are clearly better than the corresponding $HHMM(4)$. In summary, this study has shown that the application of the $PHHMM(L)$ at a typical FPR of 1% leads in most cases to an improved prediction of deleted or highly polymorphic genomic regions in C24 in comparison to the corresponding $HHMM(L)$. Subsequently, selected tree structures of the $PHHMM(2)$ are investigated.

8. Analysis of Arabidopsis Array-CGH Data

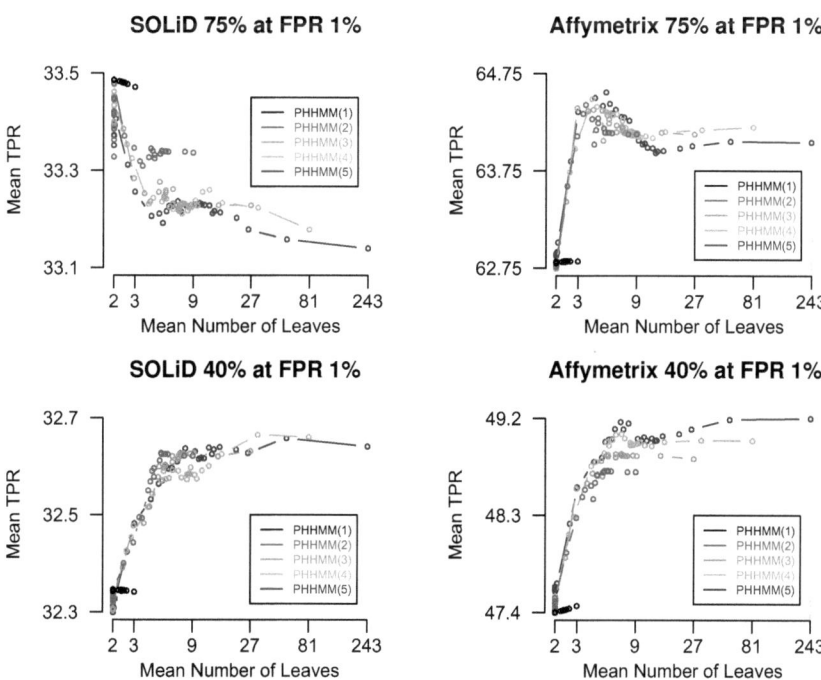

Figure 8.6: Overview of the mean TPRs obtained at a fixed FPR of 1% for the different initializations of each *PHHMM(L)* with initial order L on the Array-CGH data set under consideration of the SOLiD and the Affymetrix validation data. The mean number of leaves quantifies the complexity of each model for its twenty different initializations. The top graphics represent the results obtained for the stringent coverage of 75%, and the bottom graphics that for the less stringent coverage of 40%. For each *PHHMM(L)* the corresponding *HHMM(L)* with the highest model complexity of 3^L leaves is the rightmost point of the corresponding points shown for this *PHHMM(L)*. The mean TPRs of the *HHMM(L)* at a fixed FPR of 1% are also separately shown in Fig. 8.5.

8. Analysis of Arabidopsis Array-CGH Data

8.3.5 Selected tree structures of PHHMMs

The tree structure obtained for a *PHHMM*(L) represents the equivalence classes of state contexts $i \in S^L$ of length L for transitions from the current state of the state context i to a next state $j \in S$. The *HHMM*(L) with a complete tree, which contains each state context in a single equivalence class, is a special case of the *PHHMM*(L). Some selected trees that can underlie the *PHHMM*(1) and the *PHHMM*(2) are illustrated in Fig. 4.1. The model complexity of the *HHMM*(L) is reduced by the *PHHMM*(L) based on a parsimonious representation of the state contexts. For the Array-CGH data set, an overview of selected tree structures obtained for the *PHHMM*(2) of initial order two is shown in Fig. 8.7. The *HHMM*(2) is represented by the complete tree with nine leaves. The fusion of nodes in this tree to the parsimonious tree with five leaves results in a specific *PHHMM*(2). For this model a transition from the state '+' is independent from the previous state. That means, the state '+' does no longer represent second-order transition probabilities like in the *HHMM*(2), instead it represents first-order transition probabilities like an *HMM*. Additionally, for a transition from the state '−' there is no longer a separate treatment of its previous states '+' and '−'. In analogy, state '=' does no longer differentiate between its previous states '=' or '+'. All these fusions of nodes in the complete tree structure of the *HHMM*(2) have led to a parsimonious representation that reaches the same TPR like the *HHMM*(2) on the SOLiD and the Affymetrix validation data at a fixed level of 1% FPR (TPRs: SOLiD 75%: 33.34%, SOLiD 40%: 32.62%, Affymetrix 75%: 64.14%, and Affymetrix 40%: 48.71%). An improved TPR of 64.56% on the Affymetrix 75% validation data set is reached by the *PHHMM*(2) represented by the parsimonious tree with three leaves shown in Fig. 8.7. For this model a transition from the state '−' is no longer depending on its previous state, and the states '=' and '+' of this *PHHMM*(2) share the same transition probabilities. However, in comparison to the *HHMM*(2) this *PHHMM*(2) has slightly reduced TPRs on the three other validation data sets (TPRs: SOLiD 75%: 33.31%, SOLiD 40%: 32.48%, and Affymetrix 40%: 48.58%). Subsequently, the performance of the *PHHMM*(L) is compared to that of the *HHMM*(L) for a higher FPR.

8.3.6 Comparison of PHHMMs and HHMMs at a higher FPR

The *PHHMM*(L) has initially been compared in Fig. 8.6 to the *HHMM*(L) based on the mean TPR that has been reached for a fixed FPR of 1%. The choice of a fixed FPR allows to control the number of tiles that are wrongly predicted as deleted or highly poly-

8. Analysis of Arabidopsis Array-CGH Data

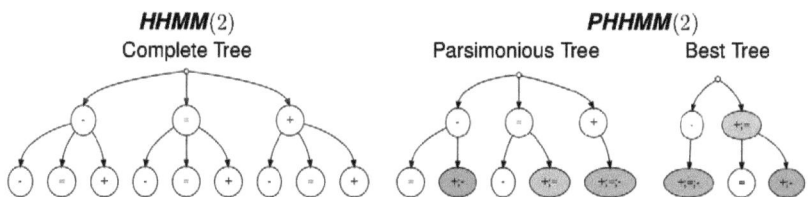

Figure 8.7: Overview of selected tree structures obtained for the *PHHMM*(2) of initial order two on the Array-CGH data set. The trees represent the equivalence classes of state contexts $i \in S^2$ of length 2 for the transition parameters. Specific fusions of nodes in the trees are colored. The *HHMM*(2) and the *PHHMM*(2) with the parsimonious tree reach both the identical performance for the SOLiD and the Affymetrix validation data sets. The *PHHMM*(2) with the shown best tree reaches the highest TPR on the Affymetrix 75% validation data set in comparison to the *HHMM*(2).

morphic in the Array-CGH data set under consideration of validation data. In this section, it is investigated how the *HHMM*(L) and the *PHHMM*(L) behave at a higher FPR of 2.5%. The results are shown in Fig. 8.8. Generally, the higher-order *PHHMM*(L) performs clearly better on all SOLiD and Affymetrix validation data sets than the first-order *PHHMM*(1). For the SOLiD 75% validation data, the third-order *PHHMM*(3) and the fifth-order *PHHMM*(5) reach a higher mean TPR than the corresponding *HHMM*(L). In addition to this, in contrast to the results obtained for this validation data at a FPR of 1% (Fig. 8.6), the higher-order *PHHMM*(L) predicts deleted or highly polymorphic regions clearly better than the first-order *PHHMM*(1) at the higher FPR of 2.5%. That means, the additional tiles that are predicted by the higher-order *PHHMM*(L) at the FPR of 2.5% are more frequently associated with a deleted or highly polymorphic region present in the SOLiD data in comparison to the additional tiles predicted by the *PHHMM*(1). For the SOLiD 40%, the Affymetrix 75%, and the Affymetrix 40% validation data, the best *PHHMM*(L) models reach a mean TPR that is comparable to that obtained by the corresponding *HHMM*(L). The advantage of the *PHHMM*(L) is that it requires a much lower model complexity to obtain this performance. That means, such a *PHHMM*(L) represents a smaller number of independent transition parameters than the corresponding *HHMM*(L). The best models for the SOLiD 40% and the Affymetrix 40% validation data are represented by the fifth-order *PHHMM*(5) within the model complexity between 27 and 81 leaves. Interestingly, only for the SOLiD 40% validation data the mean TPR of the *PHHMM*(L) increases with the order of L. For the SOLiD 75% and the Affymetrix 40% validation data, the third-order *PHHMM*(3) performs better than the fourth-order $PHHMM(4)$, and both models reach nearly the same mean TPR for the SOLiD 40% validation data. In contrast to this, the $PHHMM(4)$ performs best on the Affymetrix 75% validation data. To better characterize this model, a selected tree structure obtained among the best $PHHMM(4)$ models is shown in Fig. 8.9. The corresponding $PHHMM(4)$ still represents some specific fourth-order transition probabilities for the states '$-$' and '$=$', whereas those of state '$+$' are completely reduced to second-order transition probabilities. The TPR of 83.22% obtained for this model at 2.5% FPR is slightly better than that of the corresponding $HHMM(4)$.

In summary, at a higher FPR of 2.5% more complex models are required to reach the best performance. Generally, the third-order $PHHMM(3)$, the fourth-order $PHHMM(4)$, and the fifth-order $PHHMM(5)$ showed the best performance among all validation data sets. The mean model complexity of these best models is more shifted into the range of 9 to 81 leaves in comparison to the range of 3 to 9 leaves at a FPR of 1% in Fig. 8.6. Thus, more complex models should be preferred at this higher level of FPR. Generally, at a higher level of FPR the prediction of deleted and highly polymorphic genomic regions present in validation data becomes more and more difficult, because more and more log-ratios measured for the tiles that are predicted as deleted or highly polymorphic in the Array-CGH data set are closer to log-ratios measured for tiles that are considered as unchanged between C24 and Col based on the validation data (Fig. 8.4). These difficulties tend to be managed best by a more complex higher-order $PHHMM(L)$. Subsequently, this model is compared to other methods for analyzing Array-CGH data.

8.3.7 Comparison of PHHMMs to other methods

The $PHHMM(L)$ has initially been shown in Fig. 8.6 and Fig. 8.8 to have a good performance to predict deleted or highly polymorphic regions in the genome of C24. Here, it is investigated how this model behaves in comparison to other methods for analyzing Array-CGH data. The standard method for the analysis of the Array-CGH data set is the segMNT algorithm by Roche NimbleGen, Inc. (2008). For that reason, the predictions made by Roche NimbleGen with segMNT have been included. In addition to this, all methods summarized in Tab. 6.1 provided through the ADaCGH web-server by Diaz-Uriarte and Rueda (2007) have been tested on the Array-CGH data set. From these seven methods only ACE, CBS, FHMM, and GLAD could work with the huge number of measurements contained in the Array-CGH data set. The visual inspection of the prediction results has shown that only FHMM and GLAD are usable for the com-

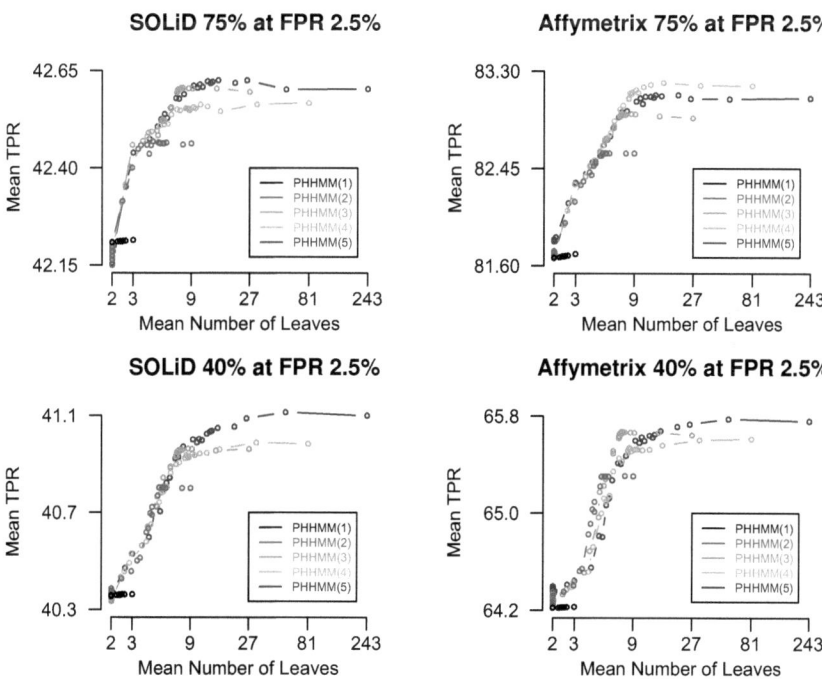

Figure 8.8: Overview of the mean TPRs obtained at a fixed FPR of 2.5% for the different initializations of the *PHHMM(L)* with initial order L on the Array-CGH data set with respect to the SOLiD and the Affymetrix validation data. The top graphics represent the results obtained for the stringent coverage of 75%, and the bottom graphics that for the less stringent coverage of 40%. The mean number of leaves quantifies the complexity of each model for its different initializations. For each *PHHMM(L)* the corresponding *HHMM(L)* with the highest model complexity of 3^L leaves is the rightmost point of the corresponding points shown for this *PHHMM(L)*. The rightmost point of the *PHHMM(1)* is the *HMM*. The results obtained at a fixed FPR of 1% are shown in Fig. 8.6.

8. Analysis of Arabidopsis Array-CGH Data

PHHMM(4): Parsimonious Tree

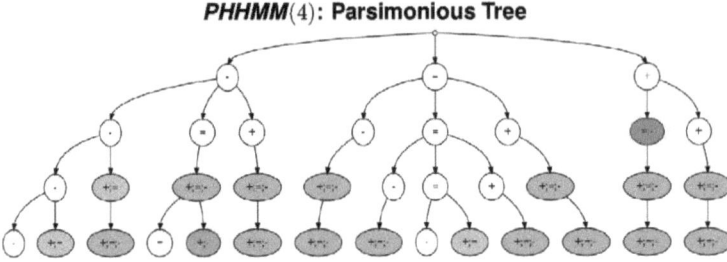

Figure 8.9: A selected tree structure among the best *PHHMM*(4) on the Affymetrix 75% validation data set shown in Fig. 8.8 at the level of 2.5% FPR. The underlying *PHHMM*(4) has a model complexity of 14 leaves, which is much less than those of the corresponding *HHMM*(4) that has a complete tree with 81 leaves. The fusions of nodes are highlighted by different colors like shown in Fig. 8.7 for the *PHHMM*(2). The states '−' and '=' still represent some of their transition probabilities as fourth-order transition probabilities, whereas for the state '+' only second-order transition probabilities remain. This *PHHMM*(4) reaches a slightly better TPR of 83.22% at 2.5% FPR than the *HHMM*(4).

parison. That means, only the predictions done by FHMM, GLAD, and segMNT could be evaluated under consideration of the SOLiD and the Affymetrix validation data sets illustrated in Fig. 8.4. All these three methods just assign one of the three states '−', '=', and '+' as label to each tile without providing a score to rank the tiles. Thus, only the point measures of TPR and FPR could be computed for each method. To compare the *PHHMM*(L) against these three methods, receiver operating characteristic (ROC) curves have been computed for the SOLiD and the Affymetrix validation data sets. For both SOLiD validation data sets a selected *PHHMM*(5) among the best models in Fig. 8.8 has been used. In analogy, for both Affymetrix validation data sets the *PHHMM*(4) with its underlying tree structure shown in Fig. 8.9 has been considered. The ROC curves including the point measures of FHMM, GLAD, and segMNT are shown in Fig. 8.10. For both types of validation data, the corresponding higher-order *PHHMM*(L) performs slightly better than GLAD. The high FPR obtained for the predictions of GLAD can be considered as a drawback for the analysis of Array-CGH data for which the majority of tiles is expected to be non-polymorphic. The comparison of the *PHHMM*(4) and the *PHHMM*(5) to FHMM and segMNT shows that the FHMM and segMNT are clearly outperformed for both types of validation data sets. The *PHHMM*(4) and the *PHHMM*(5) reach much higher TPRs at the levels of FPRs obtained for FHMM and segMNT. Generally, an additional advantage of the *PHHMM*(L)

in comparison to GLAD, FHMM, and segMNT is that the $PHHMM(L)$ provides scores to rank the tiles in Array-CGH data sets. This allows to assess the performance at each user-specified level of the FPR.

Another observation that follows from the results shown in Fig. 8.10 is that the ROC curves for SOLiD and Affymetrix validation data have clearly different shapes. This is already indicated in Fig. 8.4 that illustrates the Array-CGH data in the context of these validation data sets. Generally, higher TPRs have been obtained at fixed FPRs for the Affymetrix validation data in comparison to the SOLiD validation data. This might be caused due to the fact that the Affymetrix resequencing experiment is also based on a DNA microarray like the Array-CGH data set itself. On the other hand, the SOLiD resequencing experiment has provided much more putatively deleted or highly polymorphic genomic regions than the Affymetrix resequencing experiment. A certain proportion of these sequence polymorphisms does not seem to be present in the Array-CGH data set. This is indicated in Fig. 8.4 by log-ratios of about zero for tiles that are assumed to represent a deleted or highly polymorphic region under consideration of the SOLiD validation data. For that reason, the performance on the Array-CGH data set with respect to the SOLiD validation data is expected to be worse than that for Affymetrix validation data. However, under all methods considered here the $PHHMM(L)$ performs best on the Array-CGH data set.

8.3.8 Analysis of PHHMM predictions in the context of the genome annotation

The availability of the TAIR8 genome annotation of the reference genome of Col provides the opportunity to investigate what is functionally behind the genomic regions where the genomes of C24 and Col differ. For that reason, the $PHHMM(4)$ with the underlying parsimonious tree structure shown in Fig. 8.9 is used to predict deleted or highly polymorphic genomic regions as well as amplified genomic regions in the Array-CGH data set. Each tile in this data set has been labeled as either deleted or highly polymorphic, unchanged, or amplified by applying the State-Posterior algorithm specified in Sec. 3.4.1 with respect to the underlying three-state architecture in Fig. 8.2 of the $PHHMM(4)$. The genomic regions represented by all 17,306 tiles that have been predicted as deleted or highly polymorphic and of all 859 tiles that have been predicted as amplified have been analyzed separately in the context of their TAIR8 annotations. The obtained categorization of these tiles is shown in Fig. 8.11. By definition, the cate-

8. Analysis of Arabidopsis Array-CGH Data

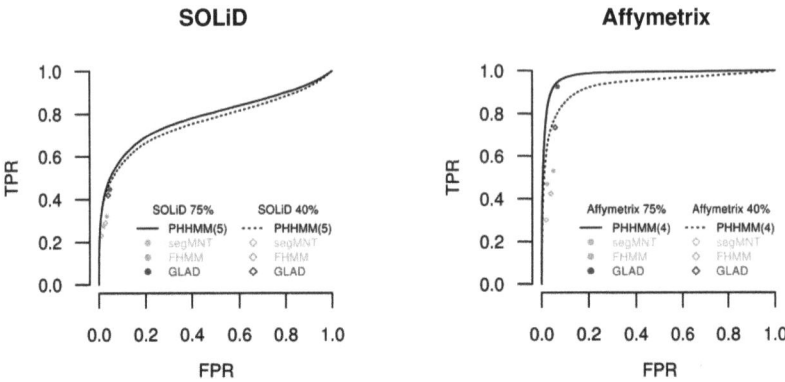

Figure 8.10: ROC curves of the *PHHMM*(4) and the *PHHMM*(5) for predicted deleted or highly polymorphic genomic regions in the Array-CGH data set under consideration of the SOLiD and the Affymetrix validation data. The point measures obtained for segMNT, FHMM, and GLAD are included to enable the comparison. The selected *PHHMM*(5) and the selected *PHHMM*(4) are among the best models shown in Fig. 8.8. The parsimonious tree structure of the *PHHMM*(4) is explicitly shown in Fig. 8.9. The ROC curves for the SOLiD and the Affymetrix validation data have clearly different shapes.

gories are not completely disjoint meaning that each tile can have annotations in more than one category. This is especially the case for tiles that are located within a gene. However, one can clearly see the trend that a large proportion of the deleted or highly polymorphic regions and of the amplified regions are caused by transposable elements. This makes sense because these mobile genomic elements might become active from time to time in the evolution of C24 and Col. Additionally, this coincides with the observation that transposable elements are overrepresented among deleted or highly polymorphic regions identified by Clark et al. (2007) for 20 accessions of *A. thaliana*. On the other hand, genes and all the categories that characterize the genes more specifically are less affected by sequence polymorphisms. Also this observation is meaningful, because if too many functional important genes would have been affected then C24 might not have been able to survive. Both observations are statistically significant in comparison to random choices of tiles highlighted by grey dashed bars in Fig. 8.11. In addition to this, all genes affected by sequence polymorphisms have functionally been categorized using the FunCat tool by Ruepp et al. (2004). The 39 genes affected by amplifications do not show a statistically significant enrichment of any functional category. In contrast to this, among the 1,675 genes affected by deleted or highly polymorphic regions four statistically significant overrepresented functional clusters of genes with p-values less than $5\cdot 10^{-6}$ have been found. The first cluster comprises 104 genes involved in ATP binding, the second cluster contains 109 genes with functions in cellular communication and signal transduction, the third cluster represents 127 genes that play a role in cell rescue, defense and virulence, and the fourth cluster contains 5 genes related to bacterial outer membrane. These findings emphasize the importance of accurate methods required by biologists for identifying sequence polymorphisms in Array-CGH data. In this chapter, the application of the $PHHMM(L)$ has shown that this model is appropriate for this task.

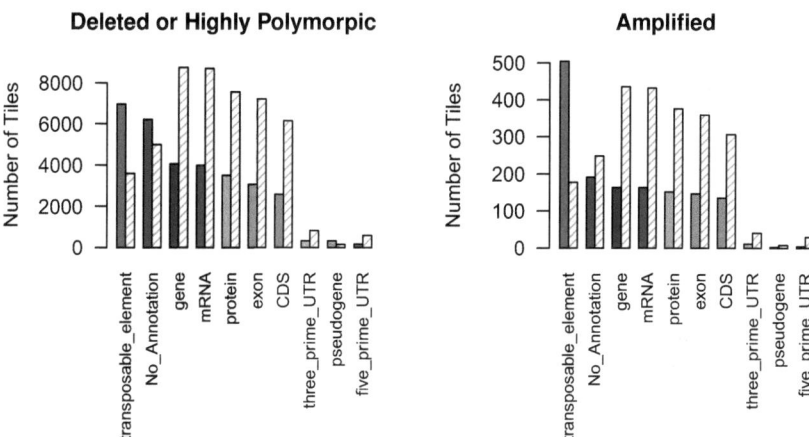

Figure 8.11: Overview of the TAIR8 genome annotations for the 17,306 tiles predicted as deleted or highly polymorphic in C24 and the 859 tiles predicted as amplified in C24 by the $PHHMM(4)$ in the Array-CGH data set using the State-Posterior algorithm. Colored bars show the counts in each category for the predictions of the $PHHMM(4)$. Grey dashed bars represent the mean values of counts for sampling 500 times the 17,306 tiles (or the 859 tiles) from the total number of tiles in the Array-CGH data set. All counts in the different categories obtained for the predictions of the $PHHMM(4)$, except 'pseudogene' for tiles predicted as amplified, differ significantly from the random counts with p-values less than 0.01.

8.4 Further reading

An initial study comparing the genomes of the Arabidopsis accessions Col and C24 has been presented at the International Conference on Bio-inspired Systems and Signal Processing (Seifert et al. (2009a)) utilizing a basic three-state first-order *HMM*. Genomic differences identified by this three-state model have been further analyzed in Banaei et al. (2011). Parts of the studies considered in this chapter and additional extensions have recently been published (Seifert et al. (2012)) including a comprehensive overview of existing Array-CGH analysis methods and related higher-order *HMM*-based approaches. A Java-based implementation of parsimonious higher-order *HMMs* is publicly available from http://www.jstacs.de/index.php/PHHMM.

9 Conclusions

The focus of this thesis has been put on extended *HMM*s and their application to recent biological high-throughput data sets generated with DNA microarrays. In the theoretical part, the algorithmic basics of standard first-order *HMM*s have been extended comprehensively to higher-order *HMM*s. Based on this sound grounding, two extensions of *HMM*s have been realized, the parsimonious higher-order *HMM* and the *HMM* with scaled transition matrices. For all these models, the integration of biological prior knowledge into the training of the model parameters has been established by the extension of the standard Baum-Welch algorithm to the Bayesian Baum-Welch algorithm. The usage of these information clearly improves the convergence to biologically meaningful model parameters. This has been demonstrated for the direct comparison of both training algorithms on the breast cancer gene expression data. Additionally, genomic features like distances between adjacent genes on a chromosome or gene pair orientations on DNA have been modeled by the usage of different transition classes. The integration of such features has led to an improved detection of differentially expressed genes in breast cancer gene expression data. Besides this, also a better identification of transcription factor target genes from ChIP-chip data has been reached. The general ability to model higher-order dependencies between adjacent measurements in the chromosomal context has been established by the development of higher-order *HMM*s. Since the number of transition parameters increases exponentially with increasing model order, the Parsimonious Cluster algorithm has been integrated into the Bayesian Baum-Welch algorithm of the higher-order *HMM* to reduce the number of independent transition parameters based on their requirement for learning the characteristics of a data set. This has led to the novel model class of parsimonious higher-order *HMM*s that includes the higher-order *HMM*s and the standard first-order *HMM*s as special cases. The modeling of higher-order dependencies by parsimonious higher-order *HMM*s has been demonstrated to improve the prediction of sequence polymorphisms in Array-CGH data. Parsimonious higher-order *HMM*s have specifically shown a better performance than higher-order *HMM*s under restrictive conditions of typically considered small false positive rates. Under less restrictive conditions, parsimonious higher-order *HMM*s have required substantially less transition parameters to reach a performance comparable to that of higher-order *HMM*s. Generally, the application of higher-order *HMM*s and parsimonious higher-order *HMM*s to the analysis of Array-CGH data has shown a clear improvement in comparison to routinely used first-order *HMM*s.

Considering the analysis of breast cancer gene expression data, the modeling of chromosomal locations and chromosomal distances of adjacent genes on DNA has led to an improved prediction of under-expressed and over-expressed genes in tumor. Based on a mixture model that ignores these additional information by modeling all gene expression levels as independent of each other, the prediction of differentially expressed genes in tumor has been much worse in comparison to the standard first-order *HMM*. The first-order *HMM* represents a natural extension of the mixture model for modeling dependencies between directly adjacent gene expression levels in the chromosomal context of the underlying genes on DNA. The usage of this *HMM* is motivated by the observation of highly positive correlations between gene expression levels in the breast cancer data set that are present due to the occurrence of amplifications and deletions of DNA segments in individual tumors. In addition to this, the tendency that two adjacent genes in close chromosomal proximity tend to have higher correlated gene expression levels than two adjacent genes in greater distance has clearly been observed for this data set. Based on that, the first-order *HMM* with scaled transition matrices has been applied to model this observation using a basic transition matrix for adjacent genes in greater chromosomal distance and another one for adjacent genes in close chromosomal proximity. The transition matrix for adjacent genes in close chromosomal proximity is computed with respect to the basic transition matrix by increasing the self-transition probabilities using a pre-defined scaling factor. Generally, the separation of adjacent genes on a chromosome into groups of genes in greater distance and genes in close chromosomal proximity, as well as the choice of a good scaling factor is depending on the data set. Since ad hoc settings are difficult, different combinations of distance thresholds and scaling factors have been tested to find biologically meaningful *HMM* parameters for the analysis of the breast cancer data set. This has led to the identification of parameter combinations for *HMM*s with scaled transition matrices that clearly improve the prediction of under-expressed and over-expressed genes in tumor in comparison to the standard first-order *HMM* that only considers chromosomal locations of genes as additional information. Thus, the prediction of differentially expressed genes in tumor has been improved stepwise by the integration of additional genomic features.

9. Conclusions

That is, the mixture model that does not include additional features has been outperformed by the standard first-order *HMM* that analyzes gene expression data in the context of chromosomal locations. This *HMM* is outperformed by the *HMM* with scaled transition matrices that additionally models chromosomal distances of genes. Genes frequently predicted as differentially expressed that have been validated independently based on data base and literature studies support these findings. Future studies could comprise gene expression data of other types of tumors; ideally data sets for which the expression status of each gene is known from independent validation experiments or annotations from domain experts. The availability of these information was limited for the considered breast cancer data set. However, such an ideal data set could be used to determine the distance threshold and the scaling factor of the *HMM* with scaled transition matrices based on a cross-validation or independent training and test data. The resulting best performing models could then be applied to follow up experiments generated on the same DNA microarray platform. Besides this, one could also further investigate the dependencies between chromosomal distances of adjacent genes and their expression levels to integrate these dependencies by a mathematical function into the self-transition probabilities of the *HMM*. Through the availability of high-density DNA microarrays, good potential is seen for the usage of higher-order *HMMs* and the development of parsimonious higher-order *HMMs* with scaled transition matrices for analyzing recent tumor expression studies.

Regarding the analysis of promoter array ChIP-chip data of the yeast *S. cerevisiae* and the model plant *A. thaliana*, the integration of chromosomal locations and chromosomal orientations of genes has improved the prediction of transcription factor target genes. The routinely used log-fold change approach that does not integrate these additional genomic features has clearly been outperformed by the standard first-order *HMM* and the *HMM* with scaled transition matrices that both make use of additional features. The standard first-order *HMM* for analyzing ChIP-chip measurements in the context of chromosomal locations of genes has been motivated through the observation of positive correlations of adjacent measurements on chromosomes. Specifically, resulting from the design of the promoter arrays, the ChIP-chip measurements of adjacent genes in head-head orientation have been observed to have much higher positive correlations than measurements of other gene pair orientations. This observation has been modeled by the usage of the *HMM* with scaled transition matrices that distinguishes between measurements of adjacent genes in head-head orientation and measurements of adjacent genes in other orientations. The specific modeling of head-head

9. Conclusions

orientations has led to a better prediction of common target genes of cell cycle specific transcription factors of the yeast *S. cerevisiae*. Moreover, a better identification of target genes of the seed-specific transcription factor ABI3 of the model plant *A. thaliana* has been reached. Generally, all predicted target genes have been validated independently based on literature and data base searches, publicly available gene expression profiles, or additional wet-lab experiments. Besides this, the same biologically motivated prior settings have been used for the promoter array ChIP-chip data of the two species. These settings have shown a good performance on these data sets indicating that end users can be supported by basic prior settings. Again, similar to the analysis of the breast cancer gene expression data, the integration of additional genomic features has led to improved models for analyzing promoter array ChIP-chip data. Initially, this has been realized by modeling dependencies between adjacent measurements on chromosomes using the first-order *HMM*. This model has been further extended by distinguishing between measurements of gene pairs in head-head orientation and other orientations using the *HMM* with scaled transition matrices. Future studies could consider the analysis of genome-wide ChIP-chip data sets based on recent high-density DNA microarrays based on higher-order *HMM*s and higher-order parsimonious *HMM*s. Since the basis of integrating additional genomic features has been established in this thesis, also the development of parsimonious higher-order *HMM*s with scaled transition matrices distinguishing between promoter and non-promoter regions on chromosomes is worth to be investigated.

Considering the analysis of the Array-CGH data set for comparing the genomes of the two accessions C24 and Columbia (Col) of *A. thaliana*, the application of higher-order *HMM*s and parsimonious higher-order *HMM*s has led to improved predictions of sequence polymorphisms in comparison to the standard first-order *HMM*. In contrast to the first-order *HMM* that only realizes dependencies between direct adjacent measurements on a chromosome, the higher-order *HMM*s extend this by modeling dependencies between a measurement and its most recent predecessors. The modeling of higher-order dependencies has been motivated by the observation of positive partial autocorrelations for the measurements along the chromosomes in the Array-CGH data set. These dependencies are present due to the fact that DNA fragments that have been extracted from an accession typically bind to several chromosomal neighboring tiles that are represented on the DNA microarray. For higher-order *HMM*s trained on the Array-CGH data set, models of order three up to five have shown the best performance for emulating these partial autocorrelations. Besides this, these models have also reached a good performance for the prediction of sequence polymorphisms in the Array-CGH data set with respect to sequence polymorphisms known from independent validation experiments. This indicates that such an initial study could also be considered for other Array-CGH data sets to determine a range of model orders that might lead to promising analysis results, but there is no direct assurance that this selection of higher-order *HMM*s leads to improved prediction results. In addition to this, most attention has been given to the ability of predicting sequence polymorphisms by parsimonious higher-order *HMM*s. These models reduce the large number of transition parameters of higher-order *HMM*s in a data-dependent manner by making use of the Parsimonious Cluster algorithm. This has enabled the modeling of sparsely higher-order dependencies between measurements in their chromosomal context. Especially at a low level of false positive predictions, the application of parsimonious higher-order *HMM*s to the Array-CGH data set has led to improved predictions of sequence polymorphisms in comparison to corresponding higher-order *HMM*s. In this context, less complex models with a clearly reduced number of independent transition parameters should be preferred. The generality of this finding still needs to be investigated in further studies with other data sets. At a greater level of false positive predictions, parsimonious higher-order *HMM*s have performed comparable or slightly better than higher-order *HMM*s. The best performing parsimonious higher-order *HMM*s have been more complex than those at a smaller level of false positive predictions. Yet, these models have clearly been less complex than corresponding higher-order *HMM*s. Generally, the validation of all these models is limited due to the diversity that has been observed for the two independent sets of sequence polymorphisms identified using the SOLiD next generation and the Affymetrix microarray-based resequencing technologies. Amplifications of DNA segments that are clearly present in the Array-CGH data set could not be validated so far. The rapid progress in the development of the next generation sequencing technologies might contribute to the establishment of better validation data for improved model validation. The comparison of results from different platforms (NimbleGen Array-CGH, SOLiD and Affymetrix resequencing) that has been started in this thesis is currently of great general interest for future studies in biology. This also includes the development of appropriate methods for analyzing the data of different platforms. In the context of Array-CGH data, the studies addressed in this thesis have shown that *HMM*s reach a better performance in comparison to other widely used approaches. Additionally, another advantage of *HMM*s is the ability to provide scores for ranking the predictions. This generally allows to assess the performance at differ-

9. Conclusions

ent levels of false positive predictions specified by the end-user. Among the *HMM*s considered for the analysis of Array-CGH data, the higher-order *HMM*s and the parsimonious higher-order *HMM*s have increased computational and memory complexities in comparison to standard first-order *HMM*s. For a standard first-order *HMM* with N states, the two basic algorithms, the Forward and the Backward algorithm, have both a computational and memory complexity of $O(T \cdot N^2)$ for a sequence of measurements of length T. This complexity is increased to $O(T \cdot N^{L+1})$ for a higher-order *HMM* of order $L > 1$. The increase in complexity is also transferred to the other algorithms of higher-order *HMM*s extended in this thesis. For the parsimonious higher-order *HMM*s that have been developed based on the algorithmic basics of higher-order *HMM*s, the computational and memory complexity of the Bayesian Baum-Welch algorithm is additionally increased due to the integration of the Parsimonious Cluster algorithm. For the analysis of the Array-CGH data set, a biologically motivated three-state architecture has been proposed and models of orders up to five have been considered. The training of a parsimonious higher-order *HMM* of order five has required about one day on a standard computer (3 GHz, 4 GB memory) in comparison to about fifteen minutes for a standard first-order *HMM*. As indicated by the great computational complexity, studies with *HMM*s of a much greater model order or with a much greater number of states are expected to be very time-consuming or even not feasible. However, the application of parsimonious higher-order *HMM*s has clearly improved the prediction of sequence polymorphisms between the two accessions C24 and Col in comparison to a standard first-order *HMM*. Thus, one should generally consider to pay the price of the greater computational complexity to obtain good predictions for further biological analyses. Such analyses also include the question what is functionally behind these genomic regions in which the genomes of C24 and Col differ. In the context of the genome annotation of Col, this has led to the identification that sequence polymorphisms between C24 and Col are mainly present due to the activity of transposons that are known as mobile genomic elements. Genes and their regulatory regions have been identified to be less affected by sequence polymorphisms. This makes sense because if too many important genes would have been affected then C24 might not have been able to survive. Future studies could investigate the activity of different transposon families and the gene expression behavior of genes affected by sequence polymorphisms. Besides this, a detailed characterization of functionally uncharacterized genomic regions affected by sequence polymorphisms should be addressed.

In summary, extended *HMM*s have been studied extensively in this thesis. The algorithmic basics of higher-order *HMM*s that include standard first-order *HMM*s as special cases have been developed in great detail. Within the scope of these basics, the integration of biological prior knowledge has been established on the basis of the Bayesian Baum-Welch algorithm. Models that have been trained using this algorithm have clearly made better predictions than models that have been trained using the standard Baum-Welch algorithm without making use of biological prior knowledge. Thus, the Bayesian Baum-Welch algorithm should be preferred for adapting *HMM*s to DNA microarray data. In addition to this finding, also the integration of additional genomic features has improved the analysis of DNA microarray data. The *HMM* with scaled transition matrices has specifically been developed to model distances between adjacent genes or to distinguish between orientations of adjacent genes on chromosomes. The application of this model to recent tumor gene expression and ChIP-chip data of different organisms including human, yeast and a model plant has led to improved predictions and demonstrates the broad usability of this concept. For the analysis of Array-CGH data, the first known application of higher-order *HMM*s has been established. Moreover, to reduce the complexity of higher-order *HMM*s the model class of parsimonious higher-order *HMM*s has been developed including the higher-order *HMM*s as special cases. The higher-order *HMM*s and especially the parsimonious higher-order *HMM*s have clearly improved the prediction of sequence polymorphisms in comparison to the standard first-order *HMM*. This indicates that parsimonious higher-order *HMM*s are appropriate for analyzing Array-CGH data. Another challenging point in this thesis has been the validation of the prediction results. Rarely, large-scale validation data sets are available for specific DNA microarray experiments. For that reason, different independent approaches including literature and data base searches, comparisons to other prediction results, or additional wet-lab experiments have been considered to evaluate the models. Generally, these different validation sources have been used successfully to identify well-suited extended *HMM*s for the analysis of recent DNA microarray data sets. According to these findings, this thesis has contributed to the investigation of extensions of *HMM*s for a broad range of applications in computational biology.

Bibliography

Abramowitz, M. and Stegun, I. A., editors (1972). *Handbook of Mathematical Functions with Formulas, Graphs, and Mathematical Tables*, volume 9. Dover Publications, Inc.

Ajmera, J. et al. (2002). Robust HMM-Based Speech/Music Segmentation. *Proc. of ICASSP, IEEE*.

Applied Biosystems (2009). SOLiD: See website of Applied Biosystems for a complete description of the platform. http://www3.appliedbiosystems.com.

Arabido-Seed (2009). A trilateral project between France, Spain, and Germany studying seed development of Arabidopsis thaliana from 2004 to 2009. http://arabidoseed.ipk-gatersleben.de.

Aycard, O., Mari, J.-F., and Washington, R. (2004). Learning to automatically detect features for mobile robots using second-order Hidden Markov Models. *Int. J. Adv. Robotic Sy.*, 1(4):231–245.

Banaei, A. M. (2009). Dynamics of chromatin modifications and other nuclear features in response to intraspecific hybridization in Arabidopsis thaliana. *PhD Thesis, Martin Luther University Halle-Wittenberg*.

Banaei, A. M., Roudier, F., Seifert, M., Bérard, C., Martin Magniette, M. L., Karimi, R., Houben, A., Colot, V., and Mette, F. M. (2011). Additive inheritance of histone modifications in Arabidopsis thaliana intraspecific hybrids. *Plant J*, 67:691 – 700.

Baum, L. E. (1972). An inequality and associated maximization technique in statistical estimation for probabilistic functions of Markov processes. *Inequalities*, 3:1–8.

Baum, L. E. and Eagen, J. A. (1967). An inequality with applications to statistical estimation for probabilistic functions of Markov processes and to model for ecology. *Bull. Amer. Math. Soc.*, 73:360–363.

Baum, L. E., Petrie, T., Soules, G., and Weiss, N. (1970). A maximization technique occuring in the statistical analysis of probabilistic functions of Markov chains. *Ann. Math. Statists.*, 41:164–171.

Becker, K. G., Barnes, K. C., Bright, T. J., and Wang, G. A. (2004). The Genetic Association Database. *Nature Genetics*, 36:431–432.

Benyoussef, L., Carincotte, C., and Derrode, S. (2008). Extension of Higher-Order HMC Modeling with Applications to Image Segmentation. *Digital Signal Processing*, 18(5):849–860.

Berchtold, A. and Raftery, A. E. (2002). The Mixture Transition Distribution Model for High-Order Markov Chains and Non-Gaussian Time Series. *Statistical Science*, 17:328–356.

Beroukhim, R. et al. (2010). The landscape of somatic copy-number alteration across human cancers. *Nature*, 463:899–905.

Bilmes, J. A. (1998). A gentle tutorial of the EM algorithm and its applications to parameter estimation for Gaussian mixture and Hidden Markov Models. *Technical Report ICSI-TR 97-021*.

Bishop, C. M. (2006). *Pattern Recognition and Machine Learning*. In Jordan, M. and Kleinberg, J. and Schölkopf, editors, Information Science and Statistics, Springer.

Blumenthal, R. et al. (2007). Expression patterns of CEACAM5 and CEACAM6 in primary and metastatic tumors. *BMC Cancer*, 7.

Bolstad, B. M., Irizarry, R. A., Astrand, M., and Speed, T. P. (2003). A comparison of normalization methods for high density oligonucleotide array data based on variance and bias. *Bioinformatics*, 19(2):185–193.

Borevitz, J. O. et al. (2003). Large-scale identification of single-feature polymorphisms in complex genomes. *Genome Research*, 13:513–523.

Bourguignon, P.-Y. and Robelin, D. (2004). Modèles de Markov parcimonieux. *Actes de JOBIM, Montréal, Canada*.

Bystroff, C. et al. (2000). HMMSTR: a Hidden Markov Model for Local Sequence-Structure Correlation in Proteins. *J. Mol. Biol.*, 301:173–190.

Callegaro, A., Basso, D., et al. (2006). A locally adaptive statistical procedure (lap) to identify differentially expressed chromosomal regions. *Bioinformatics*, 22(21):2658–2666.

Campoux, A. C. et al. (1999). Hidden Markov model approach for identifying the modular framework of the protein backbone. *Protein Engineering*, 12:1063–1073.

Caron, H. et al. (2001). The Human Transcriptome Map: Clustering of Highly Expressed Genes in Chromosomal Domains. *Science*, 291:1289–1292.

Cherry, C. (2001). A General Survey of Hidden Markov Models in Bioinformatics. http://web.cs.ualberta.ca/ colinc/projects/606project.ps.

Cherry, J. M., Ball, C., Weng, S., Juvik, S., Schmidt, R., Adler, C., Dunn, B., Dweight, S., Riles, L., Mortimer, R. K., and Botstein, D. (1997). Genetic and physical maps of Saccharomyces cerevisiae. *Nature*, 387(6632 Suppl):67–73.

Ching, W. K., Fung, E. S., and Ng, M. K. (2003). Higher-Order Hidden Markov Models with Applications to DNA Sequences. *IDEAL, LNCS 2690*, pages 535–539.

Chung, H.-R. et al. (2007). A physical model for tiling array analysis. *Bioinformatics*, 23 ISMB/ECCB:i80–i86.

Churchill, G. A. (1989). Stochastic models for heterogeneous DNA sequences. *Bull. Math. Biol.*, 51:79–94.

Clark, R. M. et al. (2007). Common Sequence Polymorphisms Shaping Genetic Diversity in Arabidopsis thaliana. *Science*, 317:338–342.

Crawley, J. J. and Furge, K. A. (2002). Identification of frequent cytogenetic aberrations in hepatocellular carcinoma using gene-expression microarray data. *Genome Biology*, 3(12).

Davidson, E. H. (2001). *Genomic Regulatory Systems*. Academic Press.

de Fonzo, V. et al. (2007). Hidden Markov Models in Bioinformatics. *Current Bioinformatics*, 2:49–61.

de Lichtenberg, U. et al. (2005). New weakly expressed cell cycle-regulated genes in yeast. *Yeast*, 22(15):1191–1201.

de Villiers, E. and du Preez, J. (2001). The advantage of using higher order HMM's for segmenting acoustic files. *Proc. of the 12th Symp. PRASA, Franschhoek, South Africa*, pages 120–122.

Dempster, A. P., Laird, N. M., and Rubin, D. B. (1977). Maximum Likelihood from Incomplete Data via the EM Algorithm. *Journal of the Royal Statistical Society B*, 39:1–38.

Depre, C. et al. (2002). H11 kinase is a novel mediator of myocardial hypertrophy in vivo. *Circulation Research*, 91:1007–1014.

Derrode, S., Carincotte, C., and Bourennane, S. (2004). Unsupervised image segmentation based on high-order hidden Markov chains. *Proc. of the International Conference on Acoustics, Speech, and Signal Processing, Montréal, Canada*, pages 769–772.

Diaz-Uriarte, R. and Rueda, O. M. (2007). ADaCGH: A Parallelized Web-Based Application and R Package for the Analysis of aCGH Data. *PLoS ONE*, 2(8):e737.

du Preez, J. A. (1998). Efficient training of high-order hidden Markov models using first-order representations. *Comput Speech Lang*, 12:23–39.

Duggan, D. J. et al. (1999). Expression profiling using cDNA microarrays. *Nature Genetics*, 21:10–14.

Durbin, R., Eddy, S., Krogh, A., and Mitchison, G. (1998). *Biological sequence analysis - Probabilistic models of proteins and nucleic acids*. Cambridge University Press.

Eddy, S. R. (1998). Profile hidden Markov models. *Bioinformatics*, 14:755–763.

Ekins, R. and Chu, F. W. (1999). Microarrays: their origins and applications. *Trends Biotechnol*, 17:217–218.

Enyenihi, A. H. and Saunders, W. S. (2003). Large-scale functional genomic analysis of sporulation and meiosis in Saccharomyces cerevisiae. *Genetics*, 163(1):47–54.

Ephraim, Y. and Merhav, N. (2002). Hidden Markov Processes. *IEEE Trans. Inform. Theory*, 48(6):1518–1569.

Evans, M., Hastings, N., and Peacock, B. (2000). *Statistical Distributions, 3rd Edition*. Wiley Series in Probability and Statistics. John Wiley & Sons, Inc.

Fan, C., Vibranovski, M. D., Chen, Y., and Long, M. (2007). A microarray based genomic hybridization method for identification of new genes in plants: Case analyses of Arabidopsis and Oryza. *J Integr Plant Biol*, 49:915–926.

Fridlyand, J., Snijders, A. M., Pinkel, D., Albertson, D. G., and Jain, A. N. (2004). Hidden Markov models approach to the analysis of array CGH data. *J. Multivariate Anal.*, 90:132–153.

Frigola, J. et al. (2006). Epigenetic remodeling in colorectal cancer results in coordinate gene suppression across an entire chromosome band. *Nature Genetics*, 38:540–549.

Galgano, M. T. et al. (2006). Comprehensive analysis of HE4 expression in normal and malignant human tissues. *Mod Pathol*, 19:847–853.

Gauvain, J.-L. and Lee, C.-H. (1991). Bayesian Learning of Gaussian Mixture Densities for Hidden Markov Models. *Proc. of the workshop on Speech and Natural Language, Pacific Grove, USA*, pages 272–277.

Gauvain, J.-L. and Lee, C.-H. (1992). Bayesian Learning for Hidden Markov Model with Gaussian Mixture State Observation Densities. *Speech Communication*, 11:205–213.

Gauvain, J.-L. and Lee, C.-H. (1994). Maximum a posteriori estimation for multivariate Gaussian mixture observations of Markov chains. *IEEE Trans. on Speech and Audio Processing*, 2:291–298.

Gestl, S. A. et al. (2002). Expression of UGT2B7, a UDP-Glucuronosyltransferase implicated in the metabolism of 4-hydroxyestone and all-trans retinoic acid, in normal human breast parenchyma and in invasive and in situ breast cancers. *AJP*, 160:1467–1479.

Giaver, G. et al. (2002). Functional profiling of the Saccharomyces cerevisiae genome. *Nature*, 418(6869):387–391.

Gohr, A. (2006). The Idea of Parsimony in Tree Based Statistical Models - Parsimonious Markov Models and Parsimonious Bayesian Networks with Applications to Classification of DNA Functional Sites. *Diploma Thesis, Martin Luther University Halle-Wittenberg*.

Gottman, J. M. (1981). *Time-Series Analysis*. Cambridge University Press.

Grau, J., Keilwagen, J., Gohr, A., Haldemann, B., Posch, S., and Grosse, I. (2012). Jstacs: A java framework for statistical analysis and classification of biological sequences. *Journal of Machine Learning Research*, 13:1967–1971.

Halsted, K. C. et al. (2008). Colagen alpha1 (X1) in normal and malignant breast tissue. *Mod Pathol*, 21:1246–1254.

He, Y. (1988). Extended Viterbi algorithm for second-order hidden Markov process. *Proc. of the IEEE 9th International Conference on Pattern Recognition, Rome, Italy*, pages 718–720.

Heidenblad, M. et al. (2005). Microarray analyses reveal strong influence of DNA copy number alterations on the transcriptional patterns in pancreatic cancer: implications for the interpretation of genomic amplifications. *Oncogene*, 24:1794–1801.

Hoheisel, J. D. (2006). Microarrays technology: beyond transcript profiling and genotype analysis. *Nature Reviews Genetics*, 7:200–210.

Horsey, E. L., Jakovljevic, J., Miles, T. D., Harnpicharnchai, P., and Wollford, J. L. (2004). Role of the yeast Rrp1 protein in the dynamics of pre-ribosome maturation. *RNA*, 10(5):813–827.

Hruz, T., Laute, O., Szabo, G., Wessendrop, F., Bleuer, S., Oertle, L., Widmayer, P., Gruissem, W., and Zimmermann, P. (2008). Genevestigator V3: A Reference Expression Database for the Meta-Analysis of Transcriptomes. *Advances in Bioinformatics*. Article ID 420747, 5 pages.

Hsu, L., Self, S. G., Grove, D., Randolph, T., Wang, K., Delrow, J. J., Loo, L., and Porter, P. (2005). Denoising array-based comparative genomic hybridization data using wavelets. *Biostatistics*, 6(2):211–226.

Humburg, P. et al. (2008). Parameter estimation for robust HMM analysis of ChIP-chip data. *BMC Bioinformatics*, 9(2).

Hupé, P., Stransky, N., Thiery, J.-P., Radvanyi, F., and Barillot, E. (2004). Analysis of array CGH data: from signal ratio to gain and loss of DNA regions. *Bioinformatics*, 20(18):3413–3422.

Hyman, E., Kauraniemi, P., et al. (2002). Impact of DNA Amplification on Gene Expression Patterns in Breast Cancer. *Cancer Research*, 62:6240–6245.

Iyer, V. R., Horak, C. E., Scafe, C. S., Botsein, D., Snyder, M., and Brown, P. O. (2001). Genomic binding sites of the yeast cell-cycle transcription factors SFB and MBF. *Nature*, 409:533–538.

Jelinek, F. (1998). *Statistical Methods for Speech Recognition*. The MIT Press.

Ji, H. and Wong, W. H. (2005). TileMap: create chromosomal map of tiling array hybridizations. *Bioinformatics*, 21:3629–3636.

Johnson, D. S. et al. (2008). Systematic evaluation of variability in ChIP-chip experiments using pre-defined DNA targets. *Genome Res*, 18:393–403.

Juang, B. H. and Rabiner, L. R. (1991). Hidden Markov Models for Speech Recognition. *Technometrics*, 33:251–272.

Jung, U. S. and Levin, D. E. (1999). Genome-wide analysis of gene expression regulated by the yeast cell wall integrity signalling pathway. *Mol Microbiol*, 34(5):1049–1057.

Junker, A., Mönke, G., Rutten, T., Keilwagen, J., Seifert, M., Thi, T. M., Renou, J. P., Balzergue, S., Viehover, P., Hähnel, U., Ludwig-Müller, J., Altschmied, L., Conrad, U., Weisshaar, B., and Bäumlein, H. (2012). Elongation-related functions of LEAFY COTYLEDON1 during the development of Arabidopsis thaliana. *Plant Journal*, 71:427–442.

Kauraniemi, P., Bärlund, M., Monni, O., and Kallioniemi, A. (2001). New Amplified and Highly Expressed Genes Discovered in the erbb2 Amplicon in Breast Cancer by cDNA Microarrays. *Cancer Research*, 61:8235–8239.

Keles, S., van der Laan, M. J., Dudoit, S., and Cawley, S. E. (2004). Multiple testing methods for ChIP-chip high density oligonucleotide array data. *Working Paper Series 147*. U.C. Berkeley Division of Biostatistics, University of California, Berkeley, CA.

Knab, B., Schliep, A., Steckemetz, B., and Wichern, B. (2003). Model-based clustering with Hidden Markov Models and its application to financial time-series data. *In M. Schader, W. Gaul, and M. Vichi, editors, Between Data Science and Applied Data Analysis, Springer*, pages 561–569.

Kriouile, A., Mari, J.-F., and Haton, J.-P. (1990). Some improvements in speech recognition based on HMM. *Proc. of the IEEE International Conference on Acoustics, Albuquerque, USA*, pages 545–548.

Krogh, A. (1994). Hidden Markov Models in Computational Biology: Applications to Protein Modeling. *J. Mol. Biol.*, 235:1501–1531.

Krogh, A. (1997). Two methods for improving performance of an HMM and their application to gene finding. *Proc. of 5-th ISMB, AAAI Press*, pages 179–186.

Kulp, D. et al. (1996). A generalized hidden Markov model for the recognition of human genes in DNA. *Proc. of 4-th ISMB, AAAI Press*, pages 134–141.

Lai, W. R., Johnson, M. D., Kucherlapati, R., and Park, P. J. (2005). Comparative analysis of algorithms for identifying amplifications and deletions in array CGH data. *Bioinformatics*, 21(19):3763–3770.

Lander, E. S. et al. (1987). Construction of multilocus genetic linkage maps in human. *PNAS*, 84:2363–2367.

Latchman, D. S. (2004). *Eukaryotic Transcription Factors.* Elsevier Academic Press, 4th edition.

Lee, C.-H., Lin, C.-H., and Juang, B.-H. (1990). A study on speaker adaptation of continuous density HMM parameters. *Proc. of the International Conference on Acoustics, Speech, and Signal Processing, Albuquerque, USA*, 1:145–148.

Lee, L.-M. and Lee, J.-C. (2006). A Study on High-Order Hidden Markov Models and Applications to Speech Recognition. *IEA/AIE, Annecy, France*, pages 682–690.

Lee, T. I., Rinaldi, N. J., Robert, F., Odom, D. T., et al. (2002). Transcripitonal Regulatory Networks in Saccaromyces cerevisiae. *Science*, 298:799–804.

Li, J. and Gray, R. M. (2000). *Image Segmentation and Compression Using Hidden Markov Models.* Kluwer Academic Publishers.

Li, W., Meyer, C. A., and Liu, X. S. (2005). A hidden Markov model for analyzing ChIP-chip experiments on genome tiling arrays and its application to p53 binding sequences. *Bioinformatics*, 21:i274–i282.

Lingjaerde, O. C., Baumbusch, L. O., Liestol, K., Glad, I. G., and Borresen-Dale, A.-L. (2005). CGH-Explorer: a program for analysis of array-CGH data. *Bioinformatics*, 21(6):821–822.

Lipshutz, R. J. et al. (1999). High density synthetic oligonucleotide arrays. *Nature Genetics*, 21:20–24.

Ma, Y. et al. (2007). Population-base molecular prognosis of breast cancer by transcriptional profiling. *Clin Cancer Research*, 13:2014–2022.

Mac Donald, I. L. and Zucchini, W. (1997). *Hidden Markov and Other Models for Discrete-valued Time Series.* Chapman & Hall.

MacKay, D. J. C. (1998). Choice of Basis for Laplace Approximation. *Machine Learning*, 33:77–86.

Maiorana, A. et al. (1995). Expression of MHC class i and class ii antigens in primary breast carcinomas and synchronous nodal metastases. *Clin Exp Meta*, 13:43–48.

Mantripragada, K. K., Buckley, P. G., de Stahl, T. D., and Dumanski, J. P. (2004). Genomic microarrays in the spotlight. *Trends Genet*, 20:87–94.

Mardis, E. R. (2008). Next-Generation DNA Sequencing Methods. *Annu. Rev. Genomics Hum. Genet.*, 9:387–402.

Mari, J.-F., Fohr, D., and Junqua, J. C. (1996). A second-order HMM for high-performance word and phoneme-based continuous speech recognition. *IEEE International Conference on Acoustics, Speech and Signal Processing, Atlanta, USA*, pages 435–438.

Mari, J.-F., Halton, J.-P., and Kriouile, A. (1997). Automatic word recognition based on second-order hidden Markov models. *IEEE Trans. on Speech and Audio Processing*, 5:22–25.

Mari, J.-F. and Haton, J.-P. (1994). Automatic word recognition based on second-order hidden Markov models. *Proc. of the 3rd International Conference on Spoken Language Processing, Yokohama, Japan*, pages 247–250.

Mari, J.-F. and Le Ber, F. (2006). Temporal and Spatial Data Mining with Second-Order Hidden Markov Models. *Soft Comput*, 10:406–414.

Marioni, J. C., Thorne, N. P., and Tavaré, S. (2006). BioHMM: a heterogeneous hidden Markov model for segmenting array CGH data. *Bioinformatics*, 22(9):1144–1146.

Martienssen, R. A., Doerge, R. W., and Colot, V. (2005). Epigenomic mapping in Arabidopsis using tiling microarrays. *Chromosome Research*, 13:299–308.

Martin-Magniette, M.-L. et al. (2008). ChIPmix: Mixture model of regressions for two-color ChIP-chip analysis. *Bioinformatics*, 24 ECCB:i181–i186.

Mc Bride, H. J., Yu, Y., and Stillman, D. J. (1999). Distinct regions of the Swi5 and Ace2 transcription factors are required for specific gene activation. *J Biol Chem*, 274(30):21029–21036.

Mönke, G., Altschmied, L., Tewes, A., Reidt, W., Mock, H. P., Bäumlein, H., and Conrad, U. (2004). Seed-specific transcription factors ABI3 and FUS3: molecular interaction with DNA. *Planta*, 219(1):158–166.

Mönke, G., Seifert, M., Keilwagen, J., Mohr, M., Grosse, I., Hähnel, U., Junker, A., Weisshaar, B., Conrad, U., Bäumlein, H., and Altschmied, L. (2012). Towards the identification and regulation of the Arabidopsis thaliana ABI3-regulon. *Nucleic Acids Research*, 40:8240–8254.

Myers, C. L., Dunham, M. J., Kung, S. Y., and Troyanskaya, O. G. (2004). Accurate detection of aneuploidies in array CGH and gene expression microarray data. *Bioinformatics*, 20(18):3533–3543.

Niu, W., Li, Z., Zhan, W., R., I. V., and Marcotte, E. M. (2008). Mechanisms of cell cycle control revealed by a systematic and quantitative overexpression screen in S. cerevisiae. *PLoS Genet*, 4(7):e1000120.

Olshen, A. B., Venkatraman, E. S., Lucito, R., and Wigler, M. (2004). Circular binary segmentation for the analysis of array-based DNA copy number data. *Biostatistics*, 5(4):557–572.

Ondov, B. D., Varadarajan, A., Passalacqua, K. D., and Bergman, N. H. (2008). Efficient mapping of Applied Biosystems SOLiD sequence data to a reference genome for functional genomic applications. *Bioinformatics*, 24:2776–2777.

Perou, C. M. et al. (2000). Molecular portraits of human breast tumours. *Nature*, 406:747–752.

Piatetsky-Shapiro, G. and Tamayo, P. (2003). Microarray Data Mining: Facing the Challenges. *SIGKDD Explorations*, 5:1–5.

Picard, F., Robin, S., Lavielle, M., Vaisse, C., and Daudin, J.-J. (2005). A statistical approach for array CGH data analysis. *BMC Bioinformatics*, 6(27).

Pinkel, D. and Albertson, D. G. (2005). Array comparative genomic hybridization and its applications in cancer. *Nature Genetics*, 37:S11–S13.

Pitman, J. (1997). Some Probabilistic Aspects of Set Partitions. *Amer Math Monthly*, 104:201–209.

Pollack, J. R. et al. (1999). Genome-wide analysis of DNA copy-number changes using cDNA microarrays. *Nature Genetics*, 23:41–46.

Pollack, J. R., Sorlie, T., Perou, C. M., Rees, C. A., Jeffrey, S. S., Lonning, P. E., Tibshirani, R., Botstein, D., Borresen-Dale, A.-L., and Brown, P. O. (2002). Microarray analysis reveals a major direct role of DNA copy number alteration in the transcriptional program of human breast tumors. *PNAS*, 99(20):12963–12968.

Raaphorst, F. M. (2005). Deregulated expression of Polycomb-group oncogenes in human malignant lymphomas and epithelial tumors. *Human Molecular Genetics*, 14:R93–R100.

Rabiner, L. R. (1989). A Tutorial on Hidden Markov Models and Selected Applications in Speech Recognition. *Proc. IEEE*, 77:257–286.

Reidt, W., Wohlfarth, T., Ellerström, M., Czihai, A., Tewes, A., Rask, L., and Bäumlein, H. (2000). Gene regulation during late embryogenesis: the RY motif of maturation-specific gene promoters is a direct target of the FUS3 gene product. *Plant Journal*, 21(5):401–408.

Ren, B., Robert, F., Wyrick, J. J., Aparicio, O., Jennings, E. G., Simon, I., Zeitlinger, J., Schreiber, J., Hannett, N., Kanin, E., Volkert, T. L., Wilson, C. J., Bell, S. P., and Young, R. A. (2000). Genome-Wide Location and Function of DNA Binding Proteins. *Science*, 290(5500):2306–2309.

Rieger, M. A., Ebner, R., Bell, D. R., Kiessling, A., Rohayem, J., Schmitz, M., Temme, A., Rieber, E. P., and Weigle, B. (2004). Identification of a Novel Mammary-Restricted Cytochrome P450, CYP4Z1, with Overexpression in Breast Carcinoma. *Cancer Research*, 64:2357–2364.

Roche NimbleGen, Inc. (2008). A Performance Comparison of Two CGH Segmentation Analysis Algorithms: DNACopy and segMNT. http://www.nimblegen.com.

Rueda, O. M. and Diaz-Uriarte, R. (2007). Flexible and Accurate Detection of Genomic Copy-Number Changes from aCGH. *PLoS Comput Biol*, 3(6):e122.

Ruepp, A. et al. (2004). The FunCat, a functional annotation scheme for systematic classification of proteins from whole genomes. *Nucleic Acids Research*, 32:5539–5545.

Sasaki, E. et al. (2007). MGB1: Breast-specific expression of MGB1/mammaglobin: an examination of 480 tumors from various organs and clinicopathological analysis of MGB1-positive breast cancers. *Mod Pathol*, 20:208–214.

Schliep, A. et al. (2003). Using hidden Markov models to analyze expression time course data. *Bioinformatics*, 19:i255–i263.

Schliep, A. et al. (2004). Robust inference of groups in gene expression time-courses using mixtures of HMMs. *Bioinformatics*, 20:i283–i289.

Schuller, B. et al. (2003). Hidden Markov Model-Based Speech Emotion Recognition. *Proc. of ICASSP, IEEE*.

Schulze, A. et al. (2001). Navigating gene expression using microarrays - a technology review. *Nature Cell Biology*, 3:190–195.

Seifert, M. (2006). Analysing microarray data using homogeneous and inhomogeneous Hidden Markov Models. *Diploma Thesis, Martin Luther University Halle-Wittenberg*.

Seifert, M. (2010). Extensions of Hidden Markov Modles for the analysis of DNA microarray data. *PhD Thesis, Martin Luther University Halle-Wittenberg*.

Seifert, M., Banaei, A., Keilwagen, J., Mette, M. F., Houben, A., Roudier, F., Colot, V., Grosse, I., and Strickert, M. (2009a). Array-based genome comparison of Arabidopsis ecotypes using Hidden Markov Models. *Proc. of the Biosignals 2009, ISBN 978-989-8111-65-4, Porto, Portugal*, pages 3–11.

Seifert, M., Cortijo, S., Colomé-Tatché, M., Johannes, F., Roudier, F., and Colot, V. (2012b). MeDIP-HMM: genome-wide identification of distinct DNA methylation states from high-density tiling arrays. *Bioinformatics*, 28:2930–2939.

Seifert, M., Gohr, A., Strickert, M., and Grosse, I. (2012). Parsimonious Higher-Order Hidden Markov Models for Improved Array-CGH Analysis with Applications to Arabidopsis thaliana. *PLoS Comp Biol*, 8(1):e1002286.

Seifert, M., Keilwagen, J., Strickert, M., and Grosse, I. (2008). Utilizing promoter pair orientations for HMM-based analysis of ChIP-chip data. *Proc. of the GCB 2008, LNI 136, Dresden, Germany*, pages 116–127.

Seifert, M., Keilwagen, J., Strickert, M., and Grosse, I. (2009b). Utilizing gene pair orientations for HMM-based analysis of ChIP-chip data. *Bioinformatics*, 25:2118–2125.

Seifert, M., Strickert, M., Schliep, A., and Grosse, I. (2011). Exploiting prior knowledge and gene distances in the analysis of tumor expression profiles with extended Hidden Markov Models. *Bioinformatics*, 27:1645–1652.

Shendure, J. et al. (2005). Accurate multiplex polony sequencing of an evolved bacterial genome. *Science*, 309:1728–1732.

Shiu, S.-H. and Borevitz, J. O. (2008). The next generation of microarray research: applications in evolutionary and ecological genomics. *Heredity*, 100:141–149.

Stransky, N. et al. (2006). Regional copy number-independent deregulation of transcription in cancer. *Nature Genetics*, 38:1386–1396.

Suzuki, M., Ketterling, G. M., Li, Q., and McCarty, D. R. (2003). Viviparous Alters Global Gene Expression Patterns through Regulation of Abscisic Acid Signaling. *Plant Physiology*, 132:1664–1677.

Telikicherla, D., Kandasamy, K., Goel, R., Ahmed, M., Mathivanan, S., Somanathan, D. S., Subbannayya, Y., Selvan, L. D. S., Ranganathan, P., and Pandey, A. (2008). A resource of molecular alterations in breast cancer. In *Proc. of the Human Genome Meeting, Hyderabad, India*.

The Arabidopsis Initiative (2000). Analysis of the genome sequence of the flowering plant Arabidopsis thaliana. *Nature*, 408:796–815.

To, A., Valon, C., Savino, G., Guilleminot, J., Devic, M., Giraudat, J., and Parcy, F. (2006). A Network of Local and Redundant Gene Regulation Governs Arabidopsis Seed Maturation. *Plant Cell*, 18:1642–1651.

Toedling, J., Schmeier, S., et al. (2004). MACAT - microarray chromosome analysis tool. *Bioinformatics*, 21(9):2112–2113.

Tomida, S. et al. (2007). Identification of a metastasis signature and the DLX4 homeobox protein as a regulator of metastasis by combined transcriptome approach. *Oncogene*, 26:4600–4608.

Vicente-Carbajosa, J. and Carbonero, P. (2005). Seed maturation: developing an intrusive phase to accomplish a quiescent state. *Int. J. Dev. Biol.*, 49:645–651.

Viterbi, A. J. (1967). Error bounds for convolutional codes and an asymptotically optimum decoding algorithm. *IEEE Trans. Inform. Theory*, 19:260–269.

Wang, Y. (2006). The Variable-length Hidden Markov Model and Its Applications on Sequential Data Mining. Technical report, Tsinghua University, Beijing, Department of Computer Science.

Willenbrock, H. and Fridlyand, J. (2005). A comparison study: applying segmentation to array CGH data for downstream analyses. *Bioinformatics*, 21(22):4084–4091.

Xu, X. et al. (2007). Identification and characterization of a novel p42.3 gene as tumor-specific and mitosis phase-dependent expression in gastric cancer. *Oncogene*, 26:7371–7379.

Yu, K. et al. (2004). Conservation of breast cancer molecular subtypes and transcriptional patterns of tumor progression across distinct ethnic populations. *Clinical Cancer Research*, 10:5508–5517.

Yu, L. et al. (2006). A survey of essential gene function in the yeast cell division. *Mol Biol Cell*, 17(11):4736–4747.

Yuan, M. and Kendziorski, C. (2006). Hidden Markov Models for Microarray Time Course Data in Multiple Biological Conditions. *J Amer Statistical Assoc*, 101:1323–1332.

Zeller, G., Clark, R. M., Schneeberger, K., Bohlen, A., Weigel, D., and Rätsch, G. (2008). Detecting polymorphic regions in Arabidopsis thaliana with resequencing microarrays. *Genome Research*, 18:918–929.

Zimmermann, P., Hirsch-Hoffmann, M., Hennig, L., and Gruissem, W. (2004). GENEVESTIGATOR. Arabidopsis Microarray Database and Analysis Toolbox. *Plant Physiol*, 136:2621–2632.

Index

Arabidopsis Array-CGH study, 118–142
  Analysis methods, 121–124
    higher-order and parsimonious higher-order HMMs, 121–124
    related methods, 124
    three-state HMM architecture, 122
  Data analysis, 124–141
    analysis of predictions in the context of the genome annotation, 139–141
    comparison of parsimonious higher-order HMMs to other methods, 136–140
    partial autocorrelation function of Array-CGH measurements, 124–126
    performance of higher-order HMMs, 129–131
    performance of parsimonious higher-order HMMs, 130–137
    resequencing data for model evaluations, 125–129
    selected state context trees, 134–135, 138
  Data set, 121
  Further reading, 142

Baum's auxiliary function
  HHMM, 31–37
  PHHMM, 59–60
  SHMM, 76
Bayesian Baum-Welch algorithm
  HHMM, 44–50
  PHHMM, 58–70
  SHMM, 74–79
Breast cancer gene expression study, 80–101
  Analysis methods, 83–89
    other methods, 88–89
    three-state HMM, 83–86
    three-state SHMM, 86–88
  Correlation of gene expression levels as a function of gene distance, 84
  Data analysis, 89–100
    comparison of Baum-Welch and Bayesian Baum-Welch training, 89–92
    comparison of HMM, SHMM and other methods, 90–93
    direct effects of gene copy number changes on gene expression levels, 97
    effect of chromosomal distances of genes on self-transition probabilities, 92–94
    hotspots of under- and over-expression, 98–100
    in-depth comparison of HMM, SHMM and GLAD, 94–96
    influence of modeling chromosomal locations and distances of genes on the prediction of differentially expressed genes, 96–98
    performance of HMM and SHMM at fixed FPR, 97
    ROC curves HMM, SHMM and mixture model, 99
  Data set, 82–84
  Further reading, 101

Hidden Markov Models, 15–50

Index

first-order, 17–18
higher-order, 19–50
    Backward algorithm, 24–25
    Backward-Variable, 24
    Baum's auxiliary function, 31–37
    Baum-Welch algorithm, 31–41
    Bayesian Baum-Welch algorithm, 44–50
    Epsilon-Variable, 35
    Forward algorithm, 22–24
    Forward-Variable, 22
    Prior distribution, 41–43
    State-posterior decoding, 26–27
    State-Posterior-Variable, 27
    Viterbi decoding, 27–31
notation scheme, 20
Hidden Markov Models with Scaled Transition Matrices, 71–79
    Bayesian Baum-Welch algorithm, 74–79
    Baum's auxiliary function, 75–76
    parameter estimation, 76–79
    transition prior, 75
    model definition, 73–74
    scaling of transition matrices, 72–73
    scaled non-self-transition probability, 73
    scaled self-transition probability, 72
    scaled transition matrix, 73
    state duration, 72

Markov Models, 6–14
    first-order, 7–10
        homogeneous, 7–8
        inhomogeneous, 8–10
    higher-order, 11–14
        homogeneous, 12–13
        inhomogeneous, 13–14

Parameter estimation
    HHMM, 38–40
    PHHMM, 64–65
    SHMM, 76–79
Parsimonious Cluster Algorithm, 67–70
    computational complexity, 69–70
    computational scheme, 67–69
Parsimonious Hidden Markov Models, 51–70
    Bayesian Baum-Welch algorithm, 58–70
    computational scheme, 61–62
    estimation of transition parameters, 64–65
    extended state context tree, 66–67
    Parsimonious cluster algorithm, 67–70
    scoring scheme for tree structures, 62–63
    tree structure prior, 60–61
    tree-based transition prior, 60
    Baum's auxiliary function, 59–60
    model definition, 57–58
    partitions of set of states, 52–54
    computing partitions, 52–53
    number of partitions, 53
    set of partitions, 54
    tree-based representation of state contexts, 54–56
    equivalence classes of state contexts, 55
    state context tree, 54
Prior Distribution
    emission prior
        HHMM, 43
    initial state prior
        HHMM, 42
    transition prior
        HHMM, 42–43
        PHHMM, 60–61
        SHMM, 75
Promoter Array ChIP-chip study, 102–117
    Analysis methods, 106–109
        log-fold-change (LFC), 106
        two-state HMM, 106–107
        two-state SHMM, 107–109
    Correlation of ChIP-chip measure-

ments as a function of gene-pair orientations, 104
Data sets, 105–106
  arabidopsis, 105–106
  yeast, 105
Further reading, 117
Identification of Arabidopsis ABI3 target genes, 111–117
  comparison of ABI3 target gene predictions by LFC, HMM and SHMM, 113–114
  systematic study of differences between HMM and SHMM, 112–113
  validation of identified ABI3 target genes, 115–117
Identification of common target genes of yeast cell cycle regulators, 109–111
  comparison of LFC, HMM and SHMM, 109–110
  validation of identified target genes, 110–111
Venn diagrams comparing LFC, HMM and SHMM, 110, 115

State Contexts
  extended state context tree, 66–67
  set of state contexts, 12
  state context, 12
  state context tree, 54

i want morebooks!

Buy your books fast and straightforward online - at one of world's fastest growing online book stores! Environmentally sound due to Print-on-Demand technologies.

Buy your books online at
www.get-morebooks.com

Kaufen Sie Ihre Bücher schnell und unkompliziert online – auf einer der am schnellsten wachsenden Buchhandelsplattformen weltweit! Dank Print-On-Demand umwelt- und ressourcenschonend produziert.

Bücher schneller online kaufen
www.morebooks.de

VDM Verlagsservicegesellschaft mbH
Heinrich-Böcking-Str. 6-8   Telefon: +49 681 3720 174   info@vdm-vsg.de
D - 66121 Saarbrücken   Telefax: +49 681 3720 1749   www.vdm-vsg.de

Printed by Books on Demand GmbH, Norderstedt / Germany